Biotechnology of Orchids and Other Ornamentals

CABI BIOTECHNOLOGY SERIES

Biotechnology, in particular the use of transgenic organisms, has a wide range of applications including agriculture, forestry, food and health. There is evidence that it could make a major impact in producing plants and animals that are able to resist stresses and diseases, thereby increasing food security. There is also potential to produce pharmaceuticals in plants through biotechnology, and provide foods that are nutritionally enhanced. Genetically modified organisms can also be used in cleaning up pollution and contamination. However, the application of biotechnology has raised concerns about biosafety, and it is vital to ensure that genetically modified organisms do not pose new risks to the environment or health. To understand the full potential of biotechnology and the issues that relate to it, scientists need access to information that not only provides an overview of and background to the field, but also keeps them up to date with the latest research findings.

This series, which extends the scope of CABI's successful "Biotechnology in Agriculture" series, addresses all topics relating to biotechnology including transgenic organisms, molecular analysis techniques, molecular pharming, *in vitro* culture, public opinion, economics, development and biosafety. Aimed at researchers, upper-level students and policy makers, titles in the series provide international coverage of topics related to biotechnology, including both a synthesis of facts and discussions of future research perspectives and possible solutions.

Titles Available

1. **Animal Nutrition with Transgenic Plants**
   Edited by G. Flachowsky
2. **Plant-derived Pharmaceuticals: Principles and Applications for Developing Countries**
   Edited by K.L. Hefferon
3. **Transgenic Insects: Techniques and Applications**
   Edited by M.Q. Benedict
4. **Bt Resistance: Characterization and Strategies for GM Crops Producing Bacillus thuringiensis Toxins**
   Edited by Mario Soberón, Yulin Gao and Alejandra Bravo
5. **Plant Gene Silencing: Mechanisms and Applications**
   Edited by Tamas Dalmay
6. **Ethical Tensions from New Technology: The Case of Agricultural Biotechnology**
   Edited by Harvey James
7. **GM Food Systems and their Economic Impact**
   Tatjana Brankov and Koviljko Lovre
8. **Endophyte Biotechnology: Potential for Agriculture and Pharmacology**
   Edited by Alexander Schouten
9. **Forest Genomics and Biotechnology**
   Richard Meilan and Matias Kirst
10. **Transgenic Insects: Techniques and Applications, 2nd Edition**
    Edited by M.Q. Benedict and Maxwell J. Scott
11. **Plant Omics: Advances in Big Data Biology**
    Edited by Hajime Ohyanagi, Eiji Yamamoto, Ai Kitazumi and Kentaro Yano

12. **Next-generation Sequencing and Agriculture**
Edited by Phillip Bayer and Dave Edwards

13. **Aquaculture and Fisheries Biotechnology: Genetic Approaches**
Rex A Dunham

14. **Algal Biotechnology**
Edited by Qiang Wang

15. **Genetic Engineering of Vegetable Crops**
Edited by Pritam Kalia

16. **Plant Breeding Technology: Future Trends and Challenges**
Edited by Zeba Khan

17. **Stress-Resilient Crops: Coordinated Omics-CRISPR-Nanotechnology Strategies**
Edited by Jen-Tsung Chen

18. **High Throughput Plant Metabolomics**
Edited by Jen-Tsung Chen

19. **Plant Biotechnology and Sustainable Agriculture: Bridging the Gap for Global Food Security**
Edited by Tariq Aftab

20. **Biotechnology of Orchids and Other Ornamentals**
Edited by Pablo Bolaños-Villegas and Fure-Chyi Chen

Jian-Zhi Huang *played a crucial role in the early stages of this project, as the idea to expand beyond orchids originated from him. His loss is great since he was an expert in orchid tissue culture, but the publication of this book will help disseminate precious knowledge that will allow for the continued expansion of ornamental production across Asia and the world.*

Biotechnology of Orchids and Other Ornamentals

Edited by

Pablo Bolaños-Villegas

and

Fure-Chyi Chen

CABI

CABI is a trading name of CAB International

CABI
Nosworthy Way
Wallingford
Oxfordshire OX10 8DE
UK

Tel: +44 (0)1491 832111
E-mail: info@cabi.org
Website: www.cabi.org

CABI
200 Portland Street
Boston
MA 02114
USA

Tel: +1 (617)682-9015
E-mail: cabi-nao@cabi.org

A catalogue record for this book is available from the British Library, London, UK.

ISBN-13: 9781800627871 (hardback)
9781800627888 (ePDF)
9781800627895 (ePub)

DOI: 10.1079/9781800627895.0000

Commissioning Editor: David Hemming
Editorial Assistant: Emma McCann
Production Editor: Rosie Hayden

Typeset by Straive, Pondicherry, India
Printed and bound by CPI Group (UK) Ltd, Croydon, CR0 4YY

Contents

Contributors

Pablo Bolaños-Villegas, Fabio Baudrit Agricultural Research Station, University of Costa Rica, La Garita District, Alajuela 20101, Costa Rica. Lankester Botanical Gardens, University of Costa Rica, Dulce Nombre District, Cartago 30109, Costa Rica. Faculty of Food and Agricultural Sciences, Rodrigo Facio Campus, School of Agronomy, University of Costa Rica, Montes de Oca County, San Jose 11503, Costa Rica. E-mail: pablo.bolanosvillegas@ucr.ac.cr

Jean Carlos Cardoso, Laboratory of Plant Physiology and Tissue Culture, Department of Biotechnology, Plant and Animal Production, Center for Agricultural Sciences, Federal University of Sao Carlos, Rodovia Anhanguera, Km 174, CEP 13600-970, Araras, São Paulo, Brazil. E-mail: jeancardoso@ufscar.br

Jia-Qi Chang, Taiwan Seed Improvement and Propagation Station, Ministry of Agriculture, No. 46, Xingzhong St., Taichung 4260, Taiwan

Fure-Chyi Chen, Department of Plant Industry, National Pingtung University of Science and Technology, Pingtung 91201, Taiwan. E-mail: fchen@mail.npust.edu.tw

Hui-Ling Chen, Hidden Agave Ranch, Escondido, California, USA

Shu-Yun Chen, Department of Agronomy, National Chung-Hsing University Crop Science Building, South District, Taichung 402, Taiwan. E-mail: syc@nchu.edu.tw

Su-Chiung Fang, Biotechnology Center in Southern Taiwan, Academia Sinica. Agricultural Biotechnology Research Center, Academia Sinica, No. 100, Sec. 1, Guiren 13th Rd., Guiren Dist., Tainan 711, Taiwan. E-mail: scfang@gate.sinica.edu.tw

Jian-Zhi Huang, Department of Plant Industry, National Pingtung University of Science and Technology, Pingtung 91201, Taiwan. *Deceased August 2025.*

Ya-Ling Huang, Kaohsiung District Agricultural Research and Extension Station (KDAIS), Pingtung 908126, Taiwan. E-mail: hyling@mail.kdais.gov.tw

Yu-Ling Huang, Kaohsiung District Agricultural Research and Extension Station (KDAIS), Pingtung 908126, Taiwan

Chong-Wei Lee, Department of Agronomy, National Chung-Hsing University Crop Science Building, South District, Taichung 402, Taiwan

Mei-Jiuan Li, Taiwan Seed Improvement and Propagation Station, Ministry of Agriculture, No. 46, Xingzhong St., Taichung 4260, Taiwan. E-mail: ann@tss.gov.tw

Ping Lim, Ornamental Plant Research, Altman Plants San Diego, California, USA. E-mail: pinglim8@msn.com

Hsuan-Ching Lin, Taiwan Seed Improvement and Propagation Station, Ministry of Agriculture, No. 46, Xingzhong St., Taichung City 4260, Taiwan

Yung Lun Lin, Department of Horticulture, National Chung Hsing University, No. 145, Xingda Road, South District, Taichung 402, Taiwan

I-Chun Pan, Department of Horticulture, National Chung Hsing University, No. 145, Xingda Road, South District, Taichung 402, Taiwan. Advanced Plant and Food Crop Biotechnology Center, National Chung Hsing University, Taiwan. Organic Agriculture Promotion Center, National Chung Hsing University, Taiwan. E-mail: icp@dragon.nchu.edu.tw

Ayu Oshin Yap Sinaga, Department of Horticulture, National Chung Hsing University, No. 145, Xingda Road, South District, Taichung 402, Taiwan. Department of Biology, Institut Teknologi Sumatera, Lampung, Indonesia

Jeremy Spath, Hidden Agave Ranch, Escondido, California, USA. E-mail: deepoceancurrents@gmail.com

Kanchit Thammasiri, Center for Gardening and Horticulture, Xishuangbanna Tropical Botanical Garden, Chinese Academy of Sciences, Menglun, Mengla, Yunnan 666303, China. Email: kanchitthammasiri@gmail.com

Kasi K. Viswanath, Department of Plant Industry, National Pingtung University of Science and Technology. Pingtung 91201, Taiwan

Kuang-Chao Wu, Taiwan Seed Improvement and Propagation Station, Ministry of Agriculture, No. 46, Xingzhong St., Taichung 4260, Taiwan

Wen-Hui Wu, Department of Horticulture, National Chung Hsing University, No. 145, Xingda Road, South District, Taichung 402, Taiwan

Preface

As the world faces increased urbanization, consumers feel the need to reconnect with nature and plants. Ornamental plants offer the possibility to create spaces of peace and contemplation that balance work and social pressures. Plants such as orchids, roses, and succulents offer a wide variety of colors and shapes that are exciting and vibrant, in a way that stimulates the imagination and is conducive to art, creativity, recreation, and health, but at the same time soothe and allow the design of living spaces that are quiet and pleasant. The breeding and selection of these plants in Asia, the USA, and Latin America contributes to their economies by allowing alternatives to traditional crops but also represents a connection to economies elsewhere creating a virtuous exchange link. This book offers an overview of scientific efforts in various laboratories and companies to address conservation, hybridization, selection, preservation, and mass multiplication that could support similar initiatives globally, benefiting farmers, consumers, and humanity as a whole. Please enjoy.

Pablo Bolaños-Villegas, University of Costa Rica

1 Breeding and Micropropagation of *Phalaenopsis* Orchids

Fure-Chyi Chen[1]*, Jian-Zhi Huang and Pablo Bolaños-Villegas[2,3,4]

[1]*Department of Plant Industry, National Pingtung University of Science and Technology, Pingtung, Taiwan;* [2]*Fabio Baudrit Agricultural Research Station, University of Costa Rica, La Garita District, Alajuela, Costa Rica;* [3]*Lankester Botanical Gardens, University of Costa Rica, Dulce Nombre District, Cartago, Costa Rica;* [4]*Faculty of Food and Agricultural Sciences, Rodrigo Facio Campus, School of Agronomy, University of Costa Rica, Montes de Oca County, San Jose, Costa Rica*

Abstract

The procedure for breeding *Phalaenopsis* orchids involves several key steps: setting breeding objectives, cross-hybridization, *in vitro* seed germination, cultivating seedlings through various growth stages, inducing the formation of floral spikes and flowering, selecting desirable lines, and using floral stalk node culture for micropropagation. The global orchid market features numerous commercial *Phalaenopsis* hybrids, with new cultivars continually emerging each year. Breeding goals for new hybrids now include achieving polyploidy, splashy color patterns, vivid floral fragrances, and growth of multiple flower spikes. To cut costs, high energy efficiency during production is now a priority. This efficiency has led to creating hybrids between *Phalaenopsis* and the *Vanda* alliance that can flower easily in simple greenhouses without air conditioning. This chapter outlines the processes of *in vitro* seed germination, embryo rescue, flower stalk node culture, and micropropagation of *Phalaenopsis* to provide an outline for a comprehensive breeding scheme. We hope this exercise in genetics, biotechnology, and engineering will be a useful guide for commercial orchid breeders worldwide.

Keywords: Orchid breeding, *Phalaenopsis*, moth orchid, micropropagation

1.1 Introduction

Phalaenopsis orchids are traded worldwide year-round because of their showy and long-lasting flowers. Commercial hybrids of *Phalaenopsis* are produced by tissue culture of floral stalk nodes and subsequent mass production of young plants, growing in sphagnum moss or bark mix substrates. Taiwan is the most critical country exporting all kinds of orchids, with *Phalaenopsis* hybrids the main product. During 2023 and 2024, Taiwan exported about $194 million of orchid live plants; about 80% were *Phalaenopsis*, according to the Customs Administration of the Ministry of Finance (Taiwan Orchid Grower Association, 2025). Tissue-cultured flask plantlets are also exported in large quantities to The Netherlands, Vietnam, and other countries, with sales reaching $14 million. Currently, the leading orchid exporters are Thailand, The Netherlands, Taiwan, Japan, the United States, and Vietnam (Observatory of Economic Complexity, n.d.). We can expect

*Corresponding author: fchen@mail.npust.edu.tw

 Biotechnology of Orchids and Other Ornamentals (Eds P. Bolaños-Villegas and F-C. Chen)
DOI: 10.1079/9781800627895.0001

potted *Phalaenopsis* plants to remain in the lead among all orchids in the global trade owing to their flowering ability under cool temperatures and industrial-like production that never fails to meet schedules (Sakanishi *et al.*, 1980).

1.2 Type and Color of *Phalaenopsis* Hybrids and Breeding Basics

Several groups of flower types are found in the *Phalaenopsis* market, including the standard type that has flowers that are large and white (extremely popular), pink or purple (popular during Chinese New Year festivities), spotted, yellow, or green, harlequin (waxy with spots), big lipped, and intermediate with a mix of colors. Examples of an assortment of *Phalaenopsis* flower types are in Fig. 1.1. A breeding program may resort to these cultivars to select suitable parents for cross-hybridization, such as crossing big-lipped with large colored-flower plants, then selecting for resulting novel progenies with big lips. The gene responsible for controlling the big-lip mutation is *PhAGL6 b* (Huang *et al.*, 2016), a MADS box gene that controls organ fate.

Cross-hybridization in orchids involves the fusion of male and female gametes after meiosis and mitosis. Meiosis regulates homologous DNA recombination to generate gametes with new allele combinations that form the basis of variation, evolution, and plant breeding. By analyzing sporad types resulting from meiosis I and II (Bolaños-Villegas *et al.*, 2008; 2017), one can estimate the parental fertility and the proportion of diploid sporads (e.g., unreduced gametes, or $2n$), which helps predict the eventual formation of polyploid progenies. Modern commercial cultivars are mostly tetraploid ($4n$) and some are diploid ($2n$), depending on the species or hybrid (Kuo *et al.*, 2005). To efficiently boost the proportion of unreduced gametes, we developed a simple and

Fig. 1.1. An assortment of flower colors and types of *Phalaenopsis* hybrids. A) Large standard white, *P.* Sogo Yukidian 'V3', a legendary hybrid. B) Pink or purple flowers in a new hybrid. C) A harlequin hybrid with its characteristic spots. D) A cascade-type harlequin with rows of perfectly aligned flowers on a stalk, *P.* DO Bru Q Saleha 'F5521'. E) A hybrid with candy-striped flowers and big lip, *P.* Tying Shin Gold Staff 'CYL10'. F) A big lip harlequin. G) A hybrid with delicately spotted flowers, also called 'French spots', *P.* OX Red Sesame. H) A hybrid with orange flowers, *P.* Giant Lin Duo Er-Guen 'GL131'; I) A miniature type hybrid, *P.* Sogo Vivien 'F858'. J) A miniature hybrid developed by National Pingtung University of Science and Technology, called CM169. K) A hybrid with yellow type flowers, *P.* Fuller's Sunset '2035'. L) A hybrid with medium-sized flowers and a long shelf life, *P.* Little Gem Stripes 'No. 1'. Author's own image.

affordable method that involves using regular mitotic inhibitors such as colchicine and trifluralin carefully dissolved in lanolin paste and gently applied to flower buds before meiosis (Hsu *et al.*, 2010). Then, a selected female parent is used as a putative receptor to these unreduced pollinia.

In *Phalaenopsis* breeding, certain female parents may be incompatible when crossed and may reject the pollen mass, thereby failing to set a capsule. A clever method to bypass this problem, the so-called cut-column method, consists of cutting half of the column and then placing the pollinia right into the style cavity (Chuang *et al.*, 2013). In this way, several cultivars are successfully fertilized and set fertile capsules with viable seeds. However, triploid (3*n*) cultivars may show unusual meiosis, with poor chromosome segregation, thus producing few viable gametes and leading to cross-incompatibility (Bolaños-Villegas *et al.*, 2008; Lee *et al.*, 2020). As an alternative, one can choose a suitable parent with a high proportion of dyad gametes (Bolaños-Villegas *et al.*, 2008) to pollinate these stubborn parents and obtain progeny.

1.3 Breeding for Multiple Spikes

Commercial cultivars of potted *Phalaenopsis* plants usually grow double flower spikes or even more spikes; however, in Japan, a single flower cascade spike is preferred. The number of spikes or flowers is genotype-specific and depends on vegetative growth as well as fertilizer supply, light quality and intensity, and temperature (van Tongerlo *et al.*, 2021). Parents with the following two traits must be selected: (i) two spikes and (ii) many flowers per spike (the flower cascade), and possibly other desirable traits such as high leaf number and good root growth. To produce novel hybrids, one can also adopt particular *Phalaenopsis* parent species with key traits such as compact leaves and/or many spikes, such as *P. micholitzii* and *P. tetraspis* (Fig. 1.2A, B, C). In these two species, the old flower spikes remain green and viable, continually growing new flower buds, thus hypothetically allowing old and new spikes to coexist in the same plant. To test this hypothesis, we crossed *P. micholitzii* with one very productive cultivar, *P.* Unimax Elfin: many seedlings produced double spikes or single spikes with branched inflorescence (Fig. 1.2D). This interspecific hybrid is called *P. Unimax Elfin* x *P. micholitzii* (Fig. 1.2E), and it produces many spikes (both old and new), so we consider it a potential bridge to cross with other cultivars. One hybrid of *P. tetraspis* (with about 25% of its genetic background), called progeny CS973 (Fig. 1.2F), also showed five spikes, which suggests the potential of multi-spike production with this species.

Fig. 1.2. *Phalaenopsis micholitzii*, *P. tetraspis*, and their interspecific hybrid as a source of multiple spikes. A) *P. micholitzii* flower. B) Spikes and plant architecture of *P. micholitzii*. C) *P. tetraspis* spikes and flowers. D) Progeny of a cross between *P. micholitzii* and *P.* Unimax Elfin, in a 6-cm pot with two spikes. E) Interspecific hybrid of *P. micholitzii* and *P. tetraspis* = *P.* Tzu Chiang Tetralitz. F) CS973, a progeny with about 25% genetic background of *P. tetraspis*, produced five spikes. Author's own image.

1.4 Breeding for Fragrance

Flowers of most commercial *Phalaenopsis* cultivars lack fragrance, with very few exceptions. Fragrance may become an essential trait for new cultivars released to the market. Yeh *et al.* (2021) reported breeding fragrant *Phalaenopsis* by analyzing scented or non-scented parents and their progenies. The authors showed that the progeny can inherit the fragrance trait. Floral scent was revealed as a complicated mixture of volatiles with about 208 different molecules in *Phalaenopsis* species and cultivars (Chen *et al.*, 2024). The volatile compounds linalool and geraniol, belonging to monoterpenoids, are probably the main components in the strong and pleasant fragrant species *P. bellina* and *P. violacea* (Hsiao *et al.*, 2006, 2008a; Chen *et al.*, 2024). The fragrance emission usually occurs in the morning under light and was found regulated by the circadian rhythm (Chuang *et al.*, 2017). From the above reports, fragrance seems to be a quantitative trait, and precise breeding is rather challenging, especially given the polyploid nature of commercial cultivars.

Suitable fragrant species or cultivars are required for practical purposes; that is, they add commercial appeal to a breeding program. For instance, accessions with fragrance for miniature breeding are easier to acquire from the market than the large standard type ones. However, one recent release of large pink/purple cultivars was available for breeding, so it may still happen. Examples of species or cultivars with fragrance are shown in Fig. 1.3. Species such as *P. amboinensis* (Fig. 1.3A), *P. bellina* (Fig. 1.3B), *P. violacea* (Fig. 1.3C), and a tetraploid *P. equestris* (Fig. 1.3D), derived from our breeding program, may be adopted depending on the purpose of a particular breeding program. The volatiles emitted by this species (*P. equestris*) include fatty acid derivatives, phenylpropanoids, and benzenoids (Hsiao *et al.*, 2006; Hsiao *et al.*, 2008a,b), although some individuals are non-scented. Another species

Fig. 1.3. Accessions of scented germplasm collected in the breeding program. A) *P. amboinensis*, a spring-to-summer flowering species with strong pleasant fragrance. B) *P. bellina* 'Chen', a superior clone with awards. C) *P. violacea* f. *indigo*, a mutant indigo colored species for so-called blue flower breeding. D) A scented tetraploid *P. equestris* obtained from diploid cross, with heavy substance and multiflora. E) *P. schilleriana* may contribute a pink color and fragrance to its progeny. F) Hybrid *P.* Caribbean Sunset. G) *P.* Charming Fortune 'Ponkan'. H) *P.* Lioulin Venus. I) *P.* Sogo Dothy 'Sogo F1898'. Author's own image.

to consider is *P. schilleriana* (Fig. 1.3E), which belongs to the *Phalaenopsis* section of the genus and has two key traits: a faint and delicate fragrance and multiple flowers on a spike. One drawback of this species is its short shelf life; however, the shelf life can be extended when it is hybridized with commercial cultivars. This species may also contribute a pink hue to large-flowered cultivars by introgression.

One hybrid with miniature flowers is *P.* Caribbean Sunset (Fig. 1.3F), which has a strong, pleasant fragrance and may be hybridized with other species or cultivars to select for novelty scented cultivars. For medium-sized flowers, one cultivar, *P.* Charming Fortune 'Ponkan', has flowers with a width of about 9 cm (Fig. 1.3G). The fragrance is subtle, so it may be a good choice for breeding orange flowers of medium size. Another medium-sized harlequin-type cultivar with yellow flowers, called *P.* Lioulin Venus (Fig. 1.3H), can be used to breed for fragrance, harlequin coloration, or a big lip. We mentioned previously that large standard-type cultivars are usually scentless. A recent release of a large pink cultivar by Sogo Orchids (Pingtung, Taiwan), called *P.* Sogo Dothy 'Sogo F1989' (Fig. 1.3I), emitted a faint scent and has been used to breed large standard cultivars with a noticeable fragrance.

1.5 Breeding for Specific Colors

Professor Mii and colleagues at Chiba University in Japan developed a transgenic blue *Phalaenopsis* called Wedding Promenade that synthesizes a delphinidin and expresses a flavonoid-3',5'-hydroxylase gene (*F3'5'H*) isolated from *Commelina communis* (Mii, 2012; Noda, 2018). This blue *Phalaenopsis* was exhibited at the 11th Asian Pacific Orchid Conference in Okinawa in 2013 to great acclaim. Then, a violet/bluish miniature called *P.* Violet Queen (Fig. 1.4A) was released later in The Netherlands. This Dutch cultivar has become widespread across the world and is now used as a parent to breed more violet hybrids.

We have used a violet/purple line derived from *P. tetraspis* (Fig. 1.4B), a second-generation progeny, with 25% parental contribution from *P. tetraspis*, to breed for a novel violet hybrid that we call CS973 (Fig. 1.4C). Another key species for breeding *Phalaenopsis* is *P. violacea* f. *indigo* (Fig. 1.4D), which originated from the HP Norton Nursery in the USA. It has since been widely used to breed very large novel blue *Phalaenopsis* hybrids.

One unusual trait of interest is a showy orange lip that many breeders are interested in. A miniature hybrid called *P.* Mini Mark has this trait (Fig. 1.4E), and the lip color is thought to

Fig. 1.4. *Phalaenopsis* hybrids and species with violet flowers are used to breed novelty colors. A) *P.* Violet Queen released years ago in the Netherlands. B) Violet flower of one *P. tetraspis* individual. C) Our CS973, a hybrid with 25% genetic background of *P. tetraspis* showing unusual violet purple flower color, bred by the first author. D) *P. violacea* f. *indigo*. E) *P.* Mini Mark with unusual orange lip. F) An orange lip mutant of *P.* Little Gem Stripes 'No. 1' derived from tissue culture. Author's own image.

derive from a recalcitrant species called *P. maculata*. Although not many registered hybrids have been derived from *P. maculata*, we expect to breed miniature hybrids with orange lips based on *P.* Mini Mark, such as mutants derived from tissue culture. For instance, *P.* Little Gem Stripes 'No. 1' has a nice orange lip (Fig. 1.4F). This hybrid grows long-lasting flowers and has been used to breed hybrids with medium-sized flowers. We expect to use *P.* Little Gem Stripes 'No. 1' for cross-hybridization with other medium-sized or miniature hybrids to introgress the orange lip trait.

1.6 Intergeneric Breeding with *Phalaenopsis* and *Vanda* Alliance Genera

Some genera from the *Vanda* alliance show a small and compact plant architecture, constant flowering, and good flowering response to cool night temperature (e.g., spike forcing). They may also be used as a bridge because they are crossable with *Phalaenopsis*, with great potential to develop novel intergeneric hybrids for the flower market. One such accession is *Holcostylis* M S Sunlight (Fig. 1.5A), a hybrid between the species *Holcoglossum flavescens* and *Rhynchostylis gigantea* (the foxtail orchid from the *Vanda* alliance), which can produce spikes about 1 or 2 months after moving to a cool room (about 25/18°C, day/night). Also, *Holcostylis* Yawi Girl (Fig. 1.5B), a hybrid between *Holcostylis* M S Sunlight and *Rhynchostylis gigantea*, can be forced to grow flower spikes under the same cool temperature regime. We have used both hybrids to cross with *Vanda miniata*, also called *Ascocentrum miniatum* (the red rust *Ascocentrum*) (Fig. 1.5C), the Japanese wind orchid *V. falcata* (日本風蘭花) (Fig. 1.5D), the velvet orchid *Luisia teres* (Fig. 1.5E), or key *Phalaenopsis* species such as *P. philippinensis* and *P. schilleriana* (Fig. 1.5F), to serve as intermediate bridges for further hybridization with other *Phalaenopsis* cultivars. One limitation for such intergeneric hybrids is the inheritance of undesirable plant architecture traits, with a tendency toward inheriting a large and clumsy *Vanda* size, and the flower size is not as large as in *Phalaenopsis*. Still, one key and noteworthy feature for these parents and hybrid progenies is a subtle and delicate fragrance during flowering. Thus, to improve plant and flower size in these intergeneric hybrids, further crosses with suitable hybrids are necessary.

Fig. 1.5. Some *Vanda* and *Phalaenopsis* alliance orchids were chosen as suitable parents to create intermediate hybrids. A) *Holcostylis* M S Sunlight. B) *H.* Yawi Girl. C) *H.* M S Sunlight x *Vanda miniata*. D) *H.* M S Sunlight x *Vanda falcata*. E) *H.* M S Sunlight x *Luisia teres*. F) *H.* M S Sunlight x *Phalaenopsis schilleriana*. Author's own image.

1.7 Micropropagation of *Phalaenopsis* Orchids

When one or several seedlings at a flowering stage show desirable commercial traits, they are selected and tagged with a number. The spike number, height, flower count, and flower size are measured, and the data and pictures are stored in a computer. The pollinia of selected seedlings can also be frozen and used for further cross-hybridization with one or several suitable parents to breed novel progenies. Often, the flower stalk can be cut with clean scissors and brought into the lab; then the stalk nodes are carefully cut for surface sterilization using diluted sodium hypochlorite solution plus one drop of a surfactant such as Tween-20. During the sterilization procedure, the stalk node can be placed in a test tube for sonication for up to 1 min to ensure thorough disinfection. Ten minutes later, the tubes with the stalk nodes are brought to the laminar flow hood to be cleansed with sterile water three times. Then the nodes are moved into a Petri dish, and both ends are cut again with a sterile knife. The stalk nodes can then be inserted into a test tube containing a suitable culture medium combining auxin and cytokinins in a desirable concentration. The whole procedure can be consulted in many publications (such as related chapters in this book) or directly with biotech companies in Taiwan. Our team has published several reports on micropropagation of *Phalaenopsis* flower stalk nodes (Wu and Chen, 2008a,b; Wu *et al.*, 2009).

Alternatively, one might resort to another strategy for micropropagation by developing protocorm-like bodies (PLBs) for the mass production of young plants. Our group has reported a protocol for the induction of PLBs from the culture of etiolated leaf segments (Hsu and Chen, 2003). PLBs can also be sectioned and then cultured in a suitable medium free of growth regulators (Huang *et al.*, 2014a). Because PLBs are equivalent to somatic embryos (Lee *et al.*, 2013), they are suitable for plant molecular studies. For instance, we studied the regulation of *SERK* genes during the multiplication of PLBs by wounding and exposure to ethylene (Huang *et al.*, 2014b).

Depending on the levels of growth regulators, the dormant axillary bud placed within the stalk node will grow into one or several new shoots after *in vitro* culture (Fig. 1.6A). After 6–8 weeks, the larger shoots in the test tube may be extracted, cut again, and cultured in a fresh medium to induce further shoot proliferation (Wu and Chen, 2008a). In the final stage of shoot multiplication, individual shoots can be excised and cultured in a medium free of growth regulators or placed in medium supplemented with auxin to promote adventitious root formation. Young plantlets at the final stage can be taken to the greenhouse, deflasked, and then cultured in a 6-cm plastic pot with wet sphagnum moss or fine bark mix, depending on the availability of potting substrates and cultivation preferences.

Eventually, careful sexual hybridization leads to the harvest of seed pods, also called capsules. These pods carry what may become valuable seedlings with valuable flower traits. These seeds require cultivation in a sterile container filled with nutrients because establishing a mycorrhizal symbiosis as in nature is not yet possible. For this purpose, we created a "mature" seed germination protocol for when the capsule turns yellow or cracks open when mature. The seeds can be dried indoors for several days, and a small amount can be sterilized with disinfectant and germinated in a nutrient medium (Fig. 1.6B). When the germinated protocorms produce roots and leaves, the young seedlings (Fig. 1.6C) can be transferred into a fresh medium for further growth until transfer to the greenhouse. One aspect affecting the efficiency of cross-hybridization is the abortion of capsules, usually 1 or 2 months after pollination. Often, it takes about 40 days for ovule differentiation to occur (Zhang and O'Neill, 1993). Therefore, sexual fertilization occurs at about 40 days after pollination. When seeds fail to develop properly before capsule collection, one must rescue the immature embryos by culturing them in a nutrient medium with high sucrose concentration to induce high osmotic pressure, an environment that resembles the ovary. One example of embryo rescue is shown in Fig. 1.6D; the capsule was harvested prematurely by about 2 months. The immature embryos on the placenta tissues can be isolated as a clump and cultured in a special rescue medium. Hence, a few hybrid seedlings may be rescued for later selection.

Fig. 1.6. Flower stalk node culture and *in vitro* seed germination of *Phalaenopsis* orchids. A) Flower stalk nodes regenerated axillary shoots after culture for weeks. B) Protocorms germinated from seed sowing after 2 to 3 weeks. C) Young *Phalaenopsis* seedlings at final stage of seed sowing and soon ready for transplanting into a suitable substrate such as sphagnum moss in a small plastic pot. D) Embryo rescue of an immature capsule from a *Phalaenopsis* cross hybridization. Author's own image.

1.8 Conclusions

Cross-hybridization for breeding *Phalaenopsis* orchids takes about 2 to 3 years from pollination to flowering. Then, the time required for clonal tissue culture, or mericloning, may take 3 to 4 years, depending on the number of young plants needed for mass commercial production. Thus, micropropagation of orchids may include tissue culture for the mass production of young plants with identical genetic makeup and *in vitro* seed germination during sexual hybridization. Each seedling has its own allele combinations due to meiotic recombination, chromosome segregation, and gamete formation after meiosis. Both breeding and micropropagation require staff with different professional training profiles and skills and perhaps knowledge of markets. Thus, governments must invest properly in vocational high schools and universities to expose students to biotechnology or horticulture programs that provide suitable training. Continued investment in research and development is also necessary for the orchid industry to stay relevant and competitive in a world in which many countries compete to breed the most beautiful and captivating orchids.

Acknowledgments

We thank our respective universities (National Pingtung University of Science and Technology, and the University of Costa Rica) for their continued funding and support of our research on orchid breeding (projects C3022, C3020, C2060 and C0244). Special thanks to Ms Laura Smales (Ontario, Canada) for editing.

References

Bolaños-Villegas, P., Chin, S.W. and Chen, F.C. (2008) Meiotic chromosome behavior and capsule setting in *Doritaenopsis* hybrids. *Journal of the American Society of Horticultural Science* 133, 107–116.

Bolaños-Villegas, P., Cascante-Amador, L., González-Lutz, M.I. and Chen, F.C. (2017) Analysis of *Phalaenopsis* bud length during meiosis to predict pollen fertility. *Journal of International Cooperation* (Taiwan) 12, 1–14.

Chen, J., Zhu, X., Zheng, R., Tong, Y., Peng, Y. *et al.* (2024) Orchestrating of native *Phalaenopsis* flower scents lighted the way through artificial selective breeding partiality in the current resource utilization. *Industrial Crops and Products* 217: 118850.

Chuang, H.T., Huang K.L. and Hsu, S.T. (2013) Using cut-column pollination method to overcome breeding barriers in *Phalaenopsis* Taipei Gold 'STM'. *Journal of the Taiwan Society for Horticultural Science* 59, 103–111.

Chuang, Y.C., Lee, M.C., Chang, Y.L., Chen, W.H. and Chen, H.H. (2017) Diurnal regulation of the floral scent emission by light and circadian rhythm in the *Phalaenopsis* orchids. *Botanical Studies* 58, 1–9.

Hsiao, Y.Y., Tsai, W.C., Kuoh, C.S., Huang, T.H., Wang, H.C., Wu, T.S., Leu, Y.L., Chen, W.H. and Chen, H.H. (2006) Comparison of transcripts in *Phalaenopsis bellina* and *Phalaenopsis equestris* (Orchidaceae) flowers to deduce monoterpene biosynthesis pathway. *BMC Plant Biology* 6, 1–14.

Hsiao, Y.Y., Tsai, W.C., Chen, W.H. and Chen, H.H. (2008a) Floral scent in *Phalaenopsis*. In: Texeira da Silva, J.A. (ed.) *Floriculture, Ornamental and Plant Biotechnology: Advances and Topical Issues Volume V*. Global Science Books, Isleworth, UK, pp. 408-420.

Hsiao, Y.Y., Jeng, M.F., Tsai, W.C., Chuang, Y.C., Li, C.Y. *et al.* (2008b) A novel homodimeric geranyl diphosphate synthase from the orchid *Phalaenopsis bellina* lacking a DD (X) 2–4D motif. *The Plant Journal* 55, 719–733.

Hsu, C.C. and Chen, F.C. (2003) Plant regeneration from protocorm-like bodies induced in etiolated leaves of *Phalaenopsis aphrodite* Rchb. f. *Journal of the Chinese Society for Horticulture Science* (Taiwan) 49, 335–342.

Hsu, S.C., Cheng, T.C., Bolaños-Villegas, P., Chin, S.W. and Chen, F.C. (2010) Pollen meiotic behavior in relation to *Phalaenopsis* breeding. *Acta Horticulturae* 878, 139–144.

Huang, Y.W., Tsai, Y.J., Cheng, T.C., Chen, J.J. and Chen, F.C. (2014a) Physical wounding and ethylene-stimulated embryogenic stem cell proliferation and plantlet regeneration in protocorm-like bodies of *Phalaenopsis* orchids. *Genetics and Molecular Research* 13, 9543–9557.

Huang, Y.W., Tsai, Y.J. and Chen, F.C. (2014b) Characterization and expression analysis of somatic embryogenesis receptor-like kinase genes from *Phalaenopsis* orchid. *Genetics and Molecular Research* 13, 10690–10703.

Huang, J.Z., Lin, C.P., Cheng, T.C., Huang, Y.W., Tsai, Y.J. *et al.* (2016) The genome and transcriptome of *Phalaenopsis* yield insights into floral organ development and flowering regulation. *Peer J* 4: e2017. https://doi.org/10.7717/peerj.2017

Kuo, H.L., Ke, C.I., Chen, Y.S., Chin, S.W. and Chen, F.C. (2005) Study on root tip squash and staining techniques of *Phalaenopsis* orchids (in Chinese). *Journal of the Chinese Society of Horticultural Science* (Taiwan) 51, 339–346.

Lee, Y.I., Hsu, S.T. and Yeung, E.C. (2013) Orchid protocorm-like bodies are somatic embryos. *American Journal of Botany* 100, 2121–2131.

Lee, Y.I., Tseng, Y.F., Lee, Y.C. and Chung, M.C. (2020) Chromosome constitution and nuclear DNA content of *Phalaenopsis* hybrids. *Scientia Horticulturae* 262: 109089.

Mii, M. (2012) Development of blue *Phalaenopsis* and blue *Dahlia* (original in Japanese). *Agriculture and Horticulture* 67, 42–46.

Noda, N. (2018) Recent advances in the research and development of blue flowers. *Breeding Science* 68, 79–87.

Observatory of Economic Complexity (n.d.) Taiwan profile. https://oec.world/en/profile/country/twn

Sakanishi, Y., Imanishi, H. and Ichida, G. (1980) Effects of temperature on growth and flowering of *Phalaenopsis amabilis*. *Bulletin of the University of Osaka Prefecture* 32, 1–9.

Taiwan Orchid Grower Association (2025) Expert values of Taiwan orchids during 2023 and 2024. Available at: https://focustaiwan.tw/politics/202503070014 (accessed 18 May 2025).

van Tongerlo, E., van Ieperen, W., Dieleman, J.A. and Marcelis, L.F.M. (2021) Vegetative traits can predict flowering quality in *Phalaenopsis* orchids despite large genotypic variation in response to light and temperature. *PLOS One* 16: e0251405. DOI: 10.1371/journal.pone.0251405

Wu, H.H. and Chen, F.C. (2008a) Induction of flower stalk nodal buds of *Phalaenopsis* and *Doritaenopsis* and effect of wounding on shoot multiplication. *Journal of the Taiwan Society for Horticulture Science* 54, 67–74.

Wu, H.H. and Chen, F.C. (2008b) Effect of plant growth regulators on shoot multiplication from flower stalk nodal buds of *Phalaenopsis* and *Doritaenopsis*. *Journal of the Taiwan Society for Horticulture Science* 54, 151–159.

Wu, H.H., Teng, M.C. and Chen, F.C. (2009) Effects of supplemental chemicals and light quality on adventitious root induction of *in vitro* grown *Phalaenopsis* and *Doritaenopsis* shoots. *Journal of the Taiwan Society for Horticulture Science* 55, 127–135.

Yeh, Y.-C., Hung, C.-Y., Chang, C.-J. and Du. L.-H. (2021) Breeding of fragrant *Phalaenopsis*. In: Chen W. H. and H. H. Chen (eds) *Orchid Biotechnology IV*. World Scientific Publishing, Singapore, pp. 43–61.

Zhang, X.S. and O'Neill, S.D. (1993) Ovary and gametophyte development are coordinately regulated by auxin and ethylene following pollination. *Plant Cell* 5, 403–418.

2 Micropropagation of Ornamental Chalk *Agave*

Hui-Ling Chen and Jeremy Spath*
Hidden Agave Ranch, Escondido, California, USA

Abstract

In this report, we introduce and describe a protocol for the propagation of the chalk agave, *Agave titanota* Gentry, that has been successfully developed by us. Leaf explants excised from the meristematic region were initially cultivated on Murashige and Skoog (MS) medium supplemented with 1 mg l^{-1} naphthaleneacetic acid (NAA) and 1 mg l^{-1} thidiazuron (TDZ), resulting in the initiation of callus proliferation. Subsequent subculturing on half-strength MS medium containing 1 mg l^{-1} TDZ and 2 mg l^{-1} benzylaminopurine (BAP) over 12 weeks promoted the formation and development of shoot primordia. Shoot multiplication was optimized by transferring the cultures to MS medium supplemented with 3 mg l^{-1} BA and 0.1 mg l^{-1} NAA, with subcultures every 4 weeks. Successful rooting of the regenerated shoots was achieved on hormone-free MS medium. This protocol hopes to contribute to the preservation and responsible use of an endangered species of great commercial value.

Keywords: *Agave titanota*, chalk agave, tissue culture, cytokinins, adventitious shoot regeneration, rooting, endangered species.

2.1 Introduction

The *Agave* genus comprises succulent plants native to Mexico and extends into southern USA, Central America, northern South America, and numerous Caribbean Islands. These plants typically thrive in arid, barren regions (Spath and Moore, 2021), and this adaptive behavior is the result of their ability to fix CO_2 via the very efficient crassulacean acid metabolism (CAM), having a superficial root system, and developing succulent leaves with thick, waxy cuticle (Tamayo-Ordóñez *et al.*, 2018).

Currently, *A. tequilana* Weber, *A. fourcroydes* (henequén), and *A. angustifolia* are the more relevant species for industrial production of many products, but a wide range of *Agave species* are also used for industrial purposes such as production of distilled beverages (tequila, mezcal, and others), production of steroidal saponins, medicinal metabolites, fiber production (sisal), and alkanes with potential as biofuels (Tamayo-Ordóñez *et al.*, 2018; Bautista-Montes *et al.*, 2022)

Among the *Agave* species, *A. titanota* (also known as *rabo de león cenizo* in Spanish) is a medium-sized plant, typically growing to 50 cm to 1 m in height. It is highly favored by collectors and landscapers for its striking forms and distinctive leaf shapes with marginal teeth, making

*Corresponding author: deepoceancurrents@gmail.com

(Eds P. Bolaños-Villegas and F-C. Chen)
DOI: 10.1079/9781800627895.0002

it one of the most popular ornamental *Agave* among 310 currently accepted species of the genus *Agave* (Spath and Moore, 2021). The species *A. titanota* was first discovered by the American botanist Howard S. Gentry at elevations between 1000 and 2000 meters above sea level in a limestone canyon in Rancho Tambor, Oaxaca state, Mexico. The name "*titanota*" refers to its alabaster white color (Spath and Moore, 2021). It is currently listed as an endangered species (García-Mendoza *et al.*, 2020).

Traditionally, *Agave* plants reproduce through seeds, offsets, underground rhizomes, and adventitious buds on flower stalks. However, interspecific hybridization and seed harvesting can be time consuming, with seeds typically taking 2 to 8 months to develop and dry. Additionally, while many *Agave* ornamental hybrids bloom easily, they are often sterile, making seed production challenging (Spath and Moore, 2021).

Thus, micropropagation provides a promising alternative for enhancing *Agave* propagation. Numerous studies have explored *in vitro* propagation techniques, leading to the development of various regeneration methods for *Agave* species, showcasing significant advancements in shoot proliferation and rooting efficiency (Bautista-Montes *et al.*, 2022). Commercial production often favors direct organogenesis to stimulate axillary bud growth. For example, Binh *et al.* (1990) demonstrated that applying 0.075 mg l^{-1} naphthaleneacetic acid (NAA); 0.1 mg l^{-1} indole-3-acetic acid (IBA), and 0.5 mg l^{-1} kinetin to stolon tip explants of species *A. sisalana*, *A. fourcroydes*, and *A. cantala* resulted in successful plantlet multiplication through direct organogenesis. Similarly, Ramírez-Malagón *et al.* (2008) reported successful axillary shoot formation in several *Agave* species, including *A. tequilana*, *A. salmiana* ssp. *crassispina*, *A. duranguensis*, *A. oscura*, *A. pygmaea*, and *A. victoria-reginae*, using IBA (0.01–0.5 mg l^{-1}) and 6-benzyladenine (BA) (0.5–3 mg l^{-1}) concentrations. However, in our study, the direct organogenesis approach proved ineffective. Therefore, we adopted a regeneration method through callus induction from leaf explants. This study aimed to establish a plant regeneration protocol for *A. titanota*, contributing to the sustainable commercial cultivation of this very valuable genus.

2.2 Materials and Methods

2.2.1 Plant material and sterilization

The *A. titanota* cultivar 'WK' used in this study was sourced from a collection at Hidden Agave Ranch, Escondido, CA. The one-year-old plant was carefully cleaned under running water and allowed to air dry overnight. The leaves and roots were removed, leaving only the main body containing the meristem. The explant was dissected into 1.5 cm segments in a laminar flow hood. Surface sterilization was performed using 40% commercial sodium hypochlorite solution for 30 minutes, with a few drops of Tween-20 as a surfactant. After sterilization, the explant was rinsed only once with sterile distilled water and placed directly in half-strength Murashige and Skoog (MS) (Murashige and Skoog, 1962) medium. The explant was maintained on this medium for 5 months, during which the shoot apex began to grow. Immature leaves that developed from the meristem were excised and used for callus induction.

2.2.2 Callus induction and shoot regeneration

Leaf segments (~0.5 cm^2) from immature leaves were cultured in vessels containing 25–30 ml of MS basal medium supplemented with 3% (w/v) sucrose, 0.2% (w/v) casein hydrolysate, and 0.6% (w/v) agar (Phyto Tech Labs. #A296), 1 mg l^{-1} NAA and 1 mg l^{-1} thidiazuron (TDZ). The explants were incubated for 4 weeks to induce callus formation. Following this, the cultures were transferred to a half-strength MS medium supplemented with 2 mg l^{-1} BA and 1 mg l^{-1} TDZ. Subculturing was carried out for 12 weeks, during which adventitious shoot regeneration was observed.

2.2.3 Shoot multiplication and rooting

After shoot primordia were established, the cultures were transferred to full-strength MS medium supplemented with 3 mg l^{-1} BA and 0.1 mg l^{-1} NAA for an additional 4 to 8 weeks. This stage facilitated further shoot multiplication.

Finally, the shoots were transferred to hormone-free MS medium supplemented with 3 g l^{-1} activated charcoal to promote shoot elongation and rooting.

2.2.4 Culture conditions

All cultures were maintained under controlled environmental conditions. The cultures were incubated under an integrated LED light (Commercial Electric™, 10 W input power, equivalent to 17 W fluorescent) at an irradiance of 23 μmol m^{-2} s^{-1}, with an 18-hour photoperiod. The light source was positioned 15 cm above the cultures to ensure uniform illumination. The temperature was maintained at 23°C throughout the study.

2.3 Results and Discussion: Callus Induction and Shoot Regeneration

A. titanota is a species valued for its unique morphological characteristics, which evolve as the plant matures (Fig. 2.1A). In this study, callus formation was observed at the cut surfaces of explants cultured on MS medium supplemented with 1 mg l^{-1} TDZ and 1 mg l^{-1} NAA (data not shown). Following subculture of explants onto half-strength MS medium with 1 mg l^{-1} TDZ and 2 mg l^{-1} BA, adventitious shoot initiation was successfully achieved after 12 weeks (Fig. 2.1B). The inclusion of cytokinins significantly enhanced shoot proliferation of callus tissue, from which shoots could be regenerated.

Further multiplication of shoots was attained by transferring the explants to full-strength MS

Fig. 2.1. The plant and plantlet regeneration indirectly through callus in *A. titanota* 'WK'. A) Mature plant showing the typic morphological traits of *A. titanota* cultivar 'WK'. B) Shoot primordia and callus derived from leaf explants cultured on half-strength MS medium supplemented with 1 mg l^{-1} TDZ and 2 mg l^{-1} BA after 12 weeks of cultivation. C) Adventitious shoot formation and subsequent proliferation observed after transferring callus to full-strength MS medium supplemented with 3 mg l^{-1} BA and 0.1 mg l^{-1} NAA for 8 weeks. D) Adventitious root formation observed in plantlets on hormone-free MS medium supplemented with activated charcoal. Author's own image.

medium containing 3 mg l^{-1} BA and 0.1 mg l^{-1} NAA. Successful shoot proliferation was observed after 4 weeks, with multiple adventitious buds forming (Fig. 2.1C). These findings align with reported results in *A. americana*, where the combination of 0.5 mg l^{-1} NAA and 3 mg l^{-1} BA yielded a high shoot induction rate of 69% when leaf segments were used as explants (Chen *et al.*, 2014). Additionally, using meristematic tissue in indirect organogenesis, high shoot proliferation rates of *A. americana* with 0.11 μM 2,4-dichlorophenoxyacetic acid (2,4-D) and 58.7 μM BA were achieved (Lecona-Guzmán *et al.*, 2017). Similarly, meristem segments of *A. tequilana* cultured with 1.1 μM 2,4-D and 44 μM BA resulted in callus formation, which was maintained for 3 months (Valenzuela-Sánchez *et al.*, 2006). These results underscore the importance of a high cytokinin to auxin ratio in inducing shoot formation through callus tissue.

Plantlets derived from callus tissue were fully developed after being transferred to hormone-free MS medium supplemented with activated charcoal. On this medium, plantlets exhibited successful root formation and shoot elongation (Fig. 2.1D), indicating that exogenous hormones were unnecessary during the rooting phase. Similarly, mature shoots of *A. tequilana* transferred to hormone-free medium showed the first roots after 10 days, with complete development occurring 3 weeks later (Valenzuela-Sánchez *et al.*, 2006). For future large-scale commercial production, it is essential to explore different conditions to achieve uniform and rapid root formation. For instance, the rooting percentage of *A. americana* reached 90% on half-strength MS medium supplemented with 1 mg l^{-1} IBA. Furthermore, a rooting rate of 92% was achieved with half-strength MS medium containing 1.48 μM IBA and 1.61 μM NAA (Chen *et al.*, 2014).

The described protocol presents a viable option for introducing variability and serves as a useful tool for the propagation of *A. titanota*. Moreover, this is the first report of a callus culture system for this endangered species.

References

Bautista-Montes, E., Hernández-Soriano, L. and Simpson, J. (2022) Advances in the micropropagation and genetic transformation of *Agave* species. *Plants* 11, 1757. DOI: 10.3390/plants11131757

Binh, L.T., Muoi, L.T., Oanh, H.T.K., Thang, T.D. and Phong, D.T. (1990) Rapid propagation of *Agave* by *in vitro* tissue culture. *Plant Cell, Tissue and Organ Culture* 23, 67–70.

Chen, Y., Chen, X., Hu, F., Yang, H., Yue, L., Trigiano, R. and Cheng, Z.M. (2014) Micropropagation of *Agave americana*. *HortScience* 49, 320–327.

García-Mendoza, A.J., Sandoval-Gutiérrez, D., Casas, A. and Torres-García, I. (2020) *Agave titanota*: *The IUCN Red List of Threatened Species*. Available at: https://dx.doi.org/10.2305/IUCN.UK.2020-1.RLTS.T115698292A116354558.en. (accessed 17 May 2025).

Lecona-Guzmán, C.A., Reyes-Zambrano, S., Barredo-Pool, F.A., Abud-Archila, M., Montes-Molina, J.A., Rincón-Rosales, R. and Gutiérrez-Miceli, F.A. (2017) *In vitro* propagation of *Agave americana* by indirect organogenesis. *HortScience* 52, 996–999.

Murashige, T. and Skoog, F. (1962) A revised medium for rapid growth and bioassays with tobacco tissue cultures. *Physiologia Plantarum* 15, 473–479.

Ramírez-Malagón, R., Borodanenko, A., Pérez-Moreno, L., Salas-Araiza, M., Nuñez-Palenius, H.G. and Ochoa-Alejo, N. (2008) *In vitro* propagation of three *Agave* species used for liquor distillation and three for landscape. *Plant Cell Tissue and Organ Culture* 94, 201–207.

Spath, J. and Moore, J. (2021) *Agaves: Species, Cultivars and Hybrids*. Solana Succulents Books, Solana Beach, CA, USA.

Tamayo-Ordóñez, M.C., Ayil-Gutiérrez, B.A., Tamayo-Ordóñez, Y.J., Rodríguez-Zapata, L.C., Monforte-González, M. *et al.* (2018) Review and *in silico* analysis of fermentation, bioenergy, fiber, and biopolymer genes of biotechnological interest in *Agave* L. for genetic improvement and biocatalysis. *Biotechnology Progress* 34, 1314–1334. DOI: 10.1002/btpr.2689

Valenzuela-Sánchez, K.K., Juárez-Hernández, R.E., Cruz-Hernández, A., Olalde-Portugal, V., Valverde, M.E. and Paredes-López, O. (2006) Plant regeneration of *Agave tequilana* by indirect organogenesis. *In Vitro Cellular and Developmental Biology-Plant* 42, 336–340.

3 Mutation Breeding and Micropropagation of *Monstera deliciosa*

Ya-Ling Huang[1], Fure-Chyi Chen[2]* and Yu-Ling Huang[1]
[1]*Kaohsiung District Agricultural Research and Extension Station (KDAIS), Pingtung, Taiwan;* [2]*Department of Plant Industry, National Pingtung University of Science and Technology, Pingtung, Taiwan*

Abstract

Cut foliage of *Monstera deliciosa* is widely used in flower arrangements due to its long vase life and striking foliage pattern. Variegated *M. deliciosa* is especially favored in the ornamental market for its visual appeal. In this study, we examined the potential of mutagenesis to obtain variegated mutants by subjecting seeds to gamma-ray irradiation. The germination rate was reduced by increasing the irradiation dosage. The median lethal dose was between 10 to 25 Gy. Next, seeds pre-soaked for 5 days were compared to unsoaked dry seeds for their sensitivity to gamma irradiation at various dosages (10–22.5 Gy). The highest germination rate was obtained at 10 Gy for dry seeds and 12.5 Gy for soaked seeds. The lowest germination rate was observed at 22.5 Gy for both dry and soaked seeds. The soaked seeds were more tolerant of the damage caused by higher irradiation dosages. Different patterns of leaf variegation, ranging from light green to yellow-green sectors, were observed in the irradiated seedlings. Another focus was on the micropropagation techniques of the variegated *M. deliciosa mutants*. The optimal treatment for shoot proliferation was 5 mg l^{-1} 6-benzyladenine (BA) combined with 0.25 mg l^{-1} naphthaleneacetic acid (NAA), yielding the highest number of shoots per node. Rooting was most effective on medium supplemented with 1.25 mg l^{-1} NAA only, which also promoted enhanced leaf development. The presence of BA at higher concentrations generally suppressed root growth, whereas NAA encouraged both root and leaf formation. Tissue browning due to phenolic oxidation was observed in all treatments, with the highest occurrence in the control group lacking plant growth regulators. These results demonstrate that specific combinations of BA and NAA significantly influence regeneration efficiency, and the established protocol offers a reliable foundation for the large-scale clonal propagation of this high-value cultivar. The findings provide critical insight for commercial propagation and conservation of elite variegated *Monstera* germplasm.

Keywords: Ornamental foliage, *Monstera*, gamma-irradiation, LD50, germination rates, leaf variegation

3.1 Introduction

Members of the family *Araceae*, known as the aroids, are one of the most important groups of ornamental foliage plants. They are primarily native to the tropical regions of North, Central, and South America (Brickell *et al.*, 1996; Henny and Norman, 1999). *Monstera deliciosa* and *Philodendron* spp. are usually sold as ornamental foliage plants (Chen *et al.*, 2003, 2010; Will, 1985). *M. deliciosa* is also cultivated on a small scale for cut foliage production (Will, 1985).

*Corresponding author: fchen@mail.npust.edu.tw

 Biotechnology of Orchids and Other Ornamentals
(Eds P. Bolaños-Villegas and F-C. Chen)
DOI: 10.1079/9781800627895.0003

Plants with variegated foliage are more popular due to their ornamental value, such as *Caladium* (Deng and Harbaugh, 2009), *Aglaonema* (Chen and Yeh, 2007), Cuckoo pint (Rocca *et al.*, 2011), and English ivy (Kim *et al.*, 2012). However, natural variegation is rare, making artificial induction a more efficient strategy for obtaining desirable mutations.

Mutation breeding has been adopted for plant improvement to obtain novel cultivars with elite traits, such as high yield, and biotic and abiotic stress resistance (Mandal *et al.*, 2000; Perera *et al.*, 2015; Sangsiri *et al.*, 2005). Numerous mutagenesis techniques are available, including physical methods (gamma-ray, X-ray, fast neutrons) and chemical treatments (ethyl methane sulfonate (EMS), and sodium azide) (Dhakshanamoorthy *et al.*, 2011, 2015; Gallone *et al.*, 2012; Mandal *et al.*, 2000; Pathirana, 2011; Perera *et al.*, 2015; Sagan *et al.*, 1995; Sangsiri *et al.*, 2005; Watanabe *et al.*, 2007). Therefore, this study aims to investigate the potential of gamma-ray irradiation in generating novel leaf variegation patterns in this species. Additionally, the micropropagation of *M. deliciosa* remains an area requiring further exploration. In aroids, tissue culture propagation often utilizes shoot tips, axillary buds, or leaves as explants. For example, mature leaves are used in *M. acuminata*, young buds in *M. deliciosa* (Palomeque *et al.*, 2019), axillary buds in *Aglaonema* (Marinai *et al.*, 2011), and shoot tips, spathes, leaves, or seedlings in *Anthurium andraeanum* to induce callus or shoot formation (Atak and Çelik, 2012; Budiarto, 2012; Farsi *et al.*, 2012; Murillo-Gómez *et al.*, 2014). Tissue culture techniques enable the preservation of desirable traits from the mother plant and the rapid mass production of healthy, virus-free plantlets. Previous studies have shown that specific combinations of plant growth regulators can successfully induce shoot or callus formation, making tissue culture a reliable method for large-scale production. For example, in *M. acuminata*, the most effective protocol for shoot induction involved surface sterilization with a mixture of Tween-20, 20% ethanol, 2.5% NaClO, benomyl solution, and cysteine, followed by culture on Murashige and Skoog (MS) medium supplemented with 1 mg l^{-1} 6-benzyladenine (BA), 0.5 mg l^{-1} indole-3-acetic acid (IAA), and 0.1 mg l^{-1} naphthaleneacetic acid (NAA) (Palomeque *et al.*, 2021). Moreover, the use of a temporary immersion bioreactor system (SETIS™) has been shown to enhance multiplication rates compared to traditional semi-solid agar-based media for *M. deliciosa* Liebm. 'Thai Constellation' (Jing *et al.*, 2024).

Despite the commercial importance of *M. deliciosa*, reports on its mutagenesis and micropropagation remain limited. To address this gap, the present study focuses on (i) the use of gamma-ray irradiation to induce leaf variegation, and (ii) the optimization of tissue culture protocols for mass propagation in the cultivar 'Kaohsiung No.1 – Dawn', a locally bred variegated variety.

3.2 Mutation Breeding Through Gamma Ray Irradiation

3.2.1 Plant materials

Seeds of *M. deliciosa* were purchased from Asusa Spike Seeds (Taipei, Taiwan). Gamma-ray irradiation of seeds was carried out at the Institute of Nuclear Energy Research of the Atomic Energy Council, Taiwan. The irradiated seeds were germinated and grown in peat moss (BVB Substrates 70L, The Netherlands) and covered with a polyethylene-panel-covered house under 70% to 80% light shade at the Kaohsiung District Agricultural Research and Extension Station (KDAIS), in Pingtung County, Taiwan. Plants were fertigated biweekly with 1 g l^{-1} of Peters® 20N-20P_2O_5-20K_2O fertilizer (Scotts Co. LLC, USA). All plants were transferred to 15 cm pots when they grew to their sixth true leaf.

3.2.2 Seed irradiation

Seeds were irradiated at 0, 2, 5, 10, 25, 50, and 75 gray (Gy) to check for the median lethal dose (the LD_{50}). The irradiated seeds were sown in a 50-cell plastic tray with peat moss. After sowing and growing for two months, the germination rate was registered and the LD_{50} calculated.

3.2.3 Effect of seed pre-soaking on germination rate after radiation

Seeds were soaked for 5 days in water at ambient temperature before irradiation. The same amount of the fresh dry seeds was also irradiated as a control. Both sets of seeds were subjected to irradiation at doses of 10, 12.5, 15, 17.5, 20, and 22.5 Gy, respectively. The dose delivery rate was 1.0 Gy s^{-1}. After irradiation, the seeds were sown in a 50-cell plastic tray with peat moss and fertigation, in conditions mentioned previously. Two months after sowing, the germination rate was registered.

3.2.4 Results and discussion

To determine the optimal dosage of irradiation, *M. deliciosa* seeds were treated at 0, 2, 5, 10, 25, 50, and 75 Gy to check for LD_{50}. Seed germination rate was reduced by increasing the dosage. The lethal dose corresponding to 50% seed germination reduction was revealed to range between 10 to 25 Gy. As the dose increased, germination declined sharply, with significant suppression observed at 25 Gy and complete lethality at 75 Gy.

The reason could be attributed to factors such as the seed coat permeability of *M. deliciosa*, its moisture absorption dynamics, or potential protective biochemical responses activated during soaking. Among the surviving seedlings from both the dry and presoaked irradiated groups, a considerable mutation frequency was observed, ranging from 31.3% to 60.0% (data not shown). Notably, several individuals exhibited distinct leaf variegation patterns, a highly desirable trait in ornamental foliage plants. The successful development of a stable variegated line, *M. deliciosa* 'Kaohsiung No.1 – Dawn', underscores the potential of gamma irradiation as an effective tool for creating novel ornamental cultivars. This approach opens new avenues for expanding phenotypic diversity in *Monstera* and other aroids, especially in species with low natural mutation rates or limited breeding resources (Fig. 3.1).

Fig. 3.1. Gamma ray irradiation was used to induce a new ornamental cultivar—*M. deliciosa* 'Kaohsiung No.1 – Dawn'—with apparent leaf variegation. Scale bar: 5 cm. Author's own image

3.3 Micropropagation of *Monstera*

3.3.1 Plant materials and culture medium

M. deliciosa 'Kaohsiung No.1 – Dawn' is a novel variegated mutant obtained through gamma ray irradiation of *M. deliciosa* seeds, developed as part of a mutation breeding program aimed at enhancing the ornamental value. This cultivar features stable foliar variegation, making it highly desirable for the ornamental plant market. Tissue-cultured plantlets of *M. deliciosa* 'Kaohsiung No.1 – Dawn' were obtained from the floral research unit at the Kaohsiung District Agricultural Research and Extension Station. The explants used in the experiment consisted of single shoots from selected variegated seedlings, each approximately 1 cm in length, with the apical meristem removed (Fig. 3.2).

The culture medium was based on half-strength MS medium (Murashige and Skoog, 1962), supplemented with the following combinations of cytokinin (BA) (0, 5, 12.5 mg l^{-1}) and auxin (NAA) (0, 0.25, 0.5, and 1.25 mg l^{-1}) as shown in Table 3.1. The different concentrations of BA and NAA were used to evaluate their effects on the shoot proliferation of the variegated *M. deliciosa* 'Kaohsiung No.1 – Dawn'. All media contained 30 g l^{-1} sucrose and 7.8 g l^{-1} agar (Fei Kung Agar-Agar Co., Ltd., Taiwan), with the pH adjusted to 5.6–5.8 before autoclaving. The media were poured into 100 ml wide-mouth orchid culture jars and sterilized at 121°C for 20 minutes. Each treatment combination was replicated four times, with three explants per replicate. Subculturing onto fresh medium was

Fig. 3.2. Shoot of *M deliciosa* 'Kaohsiung No.1 – Dawn' used in micropropagation experiments. Author's own image.

Table 3.1. Twelve BA and NAA concentration treatments tested for *Monstera* seed germination. Author's own table.

	BA		
NAA	0 mg l^{-1}	5 mg l^{-1}	12.5 mg l^{-1}
0 mg l^{-1}	A	E	I
0.25 mg l^{-1}	B	F	J
0.5 mg l^{-1}	C	G	K
1.25 mg l^{-1}	D	H	L

performed monthly, and the explants were monitored for shoot formation, rooting, leaf development, and browning symptoms over 2 months.

3.3.2 *In vitro* shoot proliferation of *Monstera*

Auxins and cytokinins are the most commonly used plant growth regulators in tissue culture. High auxin levels favor root formation, while high cytokinin levels promote shoot induction (Teixeira da Silva *et al.*, 2015). We observed that in *Monstera*, shoot formation occurred in all treatments, regardless of whether BA and NAA were included. The most effective shoot induction was observed in the medium containing 5 mg l^{-1} BA and 0.25 mg l^{-1} NAA, yielding an average of four new shoots per explant, followed by the treatment of 12.5 mg l^{-1} BA and 0 mg l^{-1} NAA, which produced approximately three shoots per explant. In contrast, media lacking BA and with low NAA concentrations (<0.5 mg l^{-1}) showed the least success in shoot proliferation (Table 3.2).

3.3.3 Root formation

Our results suggest that auxin plays a key role in root development in *Monstera*. The best root induction was seen in the absence of BA and with NAA at 1.25 mg l^{-1}, producing over five roots per explant on average. Another effective treatment was 0.5 mg l^{-1} NAA alone. In media containing BA, root formation was limited, especially at higher BA concentrations (Table 3.2). This aligns with known interactions where high cytokinin levels inhibit root development (Chiu and Tsai, 2010).

3.3.4 Leaf development

Leaf production followed a similar trend to root growth. The most robust leaf formation occurred with 1.25 mg l^{-1} NAA and no BA, resulting in about 1.67 leaves per explant. Slightly fewer leaves were observed with 0.5 mg l^{-1} NAA and no BA. Again, media containing BA were less favorable for leaf expansion (Table 3.2).

3.3.5 Tissue browning

After two months of culture, browning of explants was observed. This is a common issue in tissue culture caused by phenolic compounds released from plant wounds, which can oxidize and spread through the medium, inhibiting growth and leading to tissue death (Chen, 2012). The highest browning rate occurred in media without BA or NAA (28%), followed by the treatment with 0.25 mg l^{-1} NAA and no BA (23%) (Table 3.2).

Table 3.2. Effect of BA and NAA combinations on shoot, root, and leaf development after 2 months of culture. Author's own table.

Treatment								
BA (mg l^{-1})	NAA (mg l^{-1})	Shoot (Number per explant)		Root (Number per explant)		Leaf (Number per explant)		Browning (%)
0	0	1.75 ± 1.1	dc*	± 1.0	cd	0.75 ± 0.6	bcde	28a
0	0.25	1.75 ± 0.8	dc	2.00 ± 1.1	bc	0.92 ± 0.5	bcde	23ab
0	0.5	1.25 ± 0.9	d	2.92 ± 3.0	b	1.33 ± 1.1	ab	13bc
0	1.25	2.67 ± 1.8	bc	5.25 ± 5.1	a	1.67 ± 1.0	a	6c
5	0	2.67 ± 1.3	bc	0.00 ± 0.0	d	0.05 ± 0.7	cdef	4c
5	0.25	4.00 ± 2.1	a	0.00 ± 0.0	d	0.08 ± 0.3	f	2c
5	0.5	2.83 ± 2.2	bc	0.00 ± 0.0	d	0.33 ± 0.7	def	3c
5	1.25	2.42 ± 1.2	bc	0.00 ± 0.0	d	1.00 ± 1.5	bc	0c
12.5	0	3.08 ± 0.9	ab	0.00 ± 0.0	d	0.00 ± 0.0	f	4c
12.5	0.25	2.67 ± 0.9	bc	0.00 ± 0.0	d	0.00 ± 0.0	f	7c
12.5	0.5	2.83 ± 1.0	bc	0.00 ± 0.0	d	0.25 ± 0.5	ef	8c
12.5	1.25	2.58 ± 1.9	bc	0.00 ± 0.0	d	0.05 ± 0.8	cdef	6c

*Means within each column followed by the same letter are not significantly different at the 5% level by the Fisher's protected LSD test. Data represent the mean ± standard error.

3.4 Conclusions

This study highlights the effectiveness of a dual strategy of mutation breeding and tissue culture for developing and propagating novel *M. deliciosa* cultivars. Gamma-ray irradiation effectively induces variegation, while tissue culture provides a scalable propagation method for elite lines. Together, these approaches may enhance the ornamental value and commercial potential of *M. deliciosa* while also presenting promising avenues for horticultural research and the floriculture industry in Taiwan and Asia.

References

Atak, C. and Çelik, O. (2012) Micropropagation of *Anthurium* spp. In: Dhal, N.K. and Sahu, S.C. (eds) *Plant Science*. InTech, Istanbul, Turkey, pp. 241–254.

Brickell, C., Cole, T. and Zuk, J.D. (1996) *Monstera*. In: Brickell, C., Cole, T. and Zuk, J.D (eds) *A-Z Encyclopedia of Garden Plants*. The Reader's Digest Association, Toronto, Canada. p. 681.

Budiarto, K. (2012) Micro propagation of several potted *Anthurium* accessions using spathe explants. *Jurnal Natur Indonesia* 11, 59–63.

Chen, T.H. (2012) *Browning and vitrification disorders in plant tissue culture (in Chinese). Taichung District Agricultural Research and Extension Station,* special publication 111.

Chen, W.L. and Yeh, D.M. (2007) Elimination of *in vitro* contamination, shoot multiplication, and *ex vitro* root of *Aglaonema*. *HortScience* 42, 629–632.

Chen, J., Henny, R.J., McConnell, D.B. and Caldwell, R.D. (2003) Gibberellic acid affects growth and flowering of *Philodendron* 'Black Cardinal'. *Plant Growth Regulators* 41, 1–6.

Chen, J., McConnell, D.B., Henny, R.J. and Norman, D.J. (2010) The foliage plant industry. *Horticultural Review* 31, 47–112.

Chiu, Y.T. and Tsai, H.S. (2010) Tissue culture study of *Saussurea involucrata* (Tianshan snow lotus). MSc thesis. Graduate Institute of Biochemical Science and Technology, Chaoyang University of Technology, Taichung, Taiwan.

Deng, Z. and Harbaugh, B.K (2009) Leaf blotching in *caladium* (Araceae) is under simple genetic control and tightly linked to vein color. *HortScience* 44, 40–43.

Dhakshanamoorthy, D., Selvaraj, R. and Chidambaram, A.L.A. (2011) Induced mutagenesis in *Jatropha curcas* L. using gamma rays and detection of DNA polymorphism through RAPD marker. *Comptes Rendus Biologies*, 334, 24–30.

Dhakshanamoorthy, D., Selvaraj, R. Chidambaram, A.L.A. (2015) Utility of RAPD marker for genetic diversity analysis in gamma rays and ethyl methane sulphonate (EMS)-treated *Jatropha curcas* plant. *Comptes Rendus Biologies* 338, 75–82.

Farsi, M., Taghavizadehk, Y.M. and Qasemiomran, V. (2012) Micropropagation of *Anthurium andreanum* cv. Terra. *African Journal of Biotechnology* 11, 13162–13166.

Gallone, A., Hunter, A. and Douglas, G.C. (2012) Radiosensitivity of *Hebe* 'Oratia Beauty' and 'Wiri Mist' irradiated *in vitro* with γ-rays from ^{60}Co. *Scientia Horticulturae* 138, 36–42.

Henny, R.J. and Norman, D.J. (1999) Gibberellic acid-induced flowering of *Syngonium podophyllum* Schott 'White Butterfly'. *HortScience* 34, 676–677.

Jing, Y., Beleski, D. and Vendrame, W. (2024) Micropropagation and acclimatization of *Monstera deliciosa* Liebm. 'Thai Constellation'. *Horticulturae* 10, 1–11.

Kim, J., Kang, S.W., Pak, C.H. and Kim, M.S. (2012) Changes in leaf variegation and coloration of English ivy and polka dot plant under various indoor light intensities. *HortTechnology* 22, 49–55.

Mandal, A.K., Chakrabarty, A.D. and Datta, S.K. (2000) Application of *in vitro* techniques in mutation breeding of *Chrysanthemum*. *Plant Cell Tissue and Organ Culture* 60, 33–38.

Marinai, T. S., Fitriani, A., Wicaksono, A. and Silva, J.A.T. (2011) Micropropagation of *Aglaonema* using axillary shoot explants. *International Journal of Basic and Applied Sciences* 11(1), 46–53.

Murashige, T. and Skoog, F. (1962) A revised medium for rapid growth and bioassays with tobacco tissue culture. *Physiologia Plantarum* 15, 473–497.

Murillo-Gómez, P. A., Naranjo, E., Callejas, R., Atehortúa, L. and Urrea, A. (2014) Micropropagation of the native species *Anthurium antioquiense* Engl. for conservation purposes. *Agronomía Colombiana* 32, 334–340.

Palomeque, N.M.C., Bertolini, V. and Donjuan, L.I. (2019) *In vitro* establishment of *Monstera acuminata* Koch and *Monstera deliciosa* Liebm. *Acta Agronomica* 68, 196–204.

Palomeque, N.M.C., Bertolini, V. and Donjuan, L.I. (2021) *In vitro* establishment: *Monstera acuminata* Koch and *Monstera deliciosa* Liebm. *Trends in Horticulture* 4, 13–21.

Pathirana, R. (2011) Plant mutation breeding in agriculture. *CABI Reviews* 1–20. DOI: 10.1079/PAVSNNR20116032

Perera, D., Barnes, D.J., Baldwin, B.S. and Reichert, N.A. (2015) Mutagenesis of *in vitro* cultures of *Miscanthus* x *giganteus* cultivar Freedom and detecting polymorphisms of regenerated plants using ISSR markers. *Industrial Crops and Products* 65, 110–116.

Rocca, N.L., Rascio, N. and Pupillo, P. (2011) Variegation in *Arum italicum* leaves, a structural-functional study. *Plant Physiology and Biochemistry* 49, 1392–1398.

Sagan, M., Morandi, D., Tarenghi, E. and Duc, G. (1995) Selection of nodulation and mycorrhizal mutants in the model plant *Medicago truncatula* (Gaertn.) after γ-ray mutagenesis. *Plant Science* 111, 63–71.

Sangsiri, C., Sorajjapinun, W. and Srinives, P. (2005) Gamma radiation induced mutations in mungbean. *Science Asia* 31, 251–255.

Teixeira da Silva, J. A., Dobránszki, J., Winarto, B. and Zeng, S. (2015) Anthurium *in vitro*: a review. *Scienta Horticulturae* 186, 266–298.

Watanabe, S., Mizoguchi, T., Aoki, K., Kubo, Y., Mori, H. *et al.* (2007) Ethylmethanesulfonate (EMS) mutagenesis of *Solanum lycopersicum* cv. Micro-Tom for large-scale mutant screens. *Plant Biotechnology* 24, 33–38.

Will, A.A. (1985) Foliage for flower arrangements from your Florida garden. *Proceedings of the Florida State Horticultural Society* 98, 349–350.

4 Cryopreservation of Orchids and Ornamental Plants

Kanchit Thammasiri
Center for Gardening and Horticulture, Xishuangbanna Tropical Botanical Garden, Chinese Academy of Sciences, Menglun, Mengla, Yunnan, China

Abstract
Plant cryopreservation involves preserving valuable species such as orchids by sampling and storing tissues away from their natural habitat, including DNA, pollen, buds, dormant buds, embryos, and seeds. Typically, storage occurs in liquid nitrogen. This process effectively halts physical and metabolic cellular activities, allowing cryopreserved materials to be recovered and grown into whole plants. Developed since 1960, plant cryopreservation employs slow cooling and vitrification principles and includes seven major steps: preculture, osmoprotection, dehydration, liquid nitrogen storage, rapid thawing, unloading, and recovery. A variety of methods have been developed, including dormant buds, slow freezing, vitrification, encapsulation–dehydration, and encapsulation–vitrification. Plant cryopreservation offers a cost-effective, space-efficient, and environmentally friendly alternative for long-term plant conservation. Recent advancements in cryopreservation in orchids such as *Vanda coerulea* and *Bletilla striata* have reduced time requirements, and involve easy processes with simple equipment, which leads to high survival rates. Thailand and China have implemented these methods for germplasm storage, and it may allow them to achieve horticultural sustainability for generations to come.

Keywords: Orchid breeding, tropical orchids, plant cryopreservation, biodiversity, *Vanda coerulea*, sustainable horticulture

4.1 Introduction

Plant cryopreservation is the preservation of plant species in the form of tiny plant materials outside natural habitats (*ex situ* conservation), including DNA, pollen, buds, dormant buds, embryonic axis, zygotic and somatic embryos, and seeds at ultra-low temperature (–130°C). But generally it is preferred to store plant materials in liquid nitrogen at –196°C, which is the temperature level that truly stops biological activities, including all biochemical reactions within cells which leads to aging (and cell death) and to genetic changes during storage (Engelmann, 2004). In some cases, such as storing seeds, storage at –20°C is more appropriate and less expensive than storage in liquid nitrogen, and buds can be stored at –4 to –80°C, but storage cannot last long. Therefore, plant cryopreservation is another option for long-term storage. Nonetheless, cryopreserved plant materials should be checked periodically.

*Corresponding author: kanchitthammasiri@gmail.com

 Biotechnology of Orchids and Other Ornamentals (Eds P. Bolaños-Villegas and F-C. Chen)
DOI: 10.1079/9781800627895.0004

Thailand is in a hot and humid tropical zone of South-east Asia with a population of about 65 million and a total land area of 513,120 km^2. Thailand is also the 13th most plant biodiverse country after Brazil, Colombia, China, Mexico, Russia, Indonesia, Venezuela, USA, Australia, India, Peru, and Malaysia (Cronquist, 1981). Unlike territories located in temperate zones, tropical ecosystems provide wider niches and can support a much larger variety of plant, animal, and microbe species. It is estimated that there are approximately 15,000 plant species in Thailand, including 3,000 species of mushrooms and fungi, 633 species of ferns, and about 1,200 species of orchids. More than 779 species of plants possess active herbal ingredients used for traditional medicines (OEPP, 1996).

Success in preserving plant germplasm by quick freezing is based on the rate of cooling (cooling rate) and the concentration of chemicals (the cryoprotectants) that prevent the formation of ice crystals within the cells (Panis *et al.*, 2005). Plant materials to be preserved must be in a suitable growth stage, type of cell groups, and a size that can be dehydrated sufficiently and efficiently. Table 4.1 shows the plant tissues and organs used for preserving plant varieties in supercool conditions and the advantages and disadvantages of preservation.

To preserve plant materials in freezing conditions, one must understand the basic mechanisms of this principle, which are: (i) slow cooling (slow freezing) and (ii) vitrification.

Table 4.1. Plant tissues that have been used for super-cool preservation tests. Author's own table.

Plant tissues	Description	Advantages	Disadvantages
Seeds	Small seeds, seeds that are resistant to dry and extreme conditions, and seeds amenable to asexual propagation	Easy preservation of orthodox seeds (those that tolerate dry conditions)	Cannot be used with large seeds that do not tolerate either cold or dry conditions (recalcitrant seeds) or asexual propagation
Pollen	Many types	Easy for different genera and helpful in plant breeding	Can store only half of the total number of chromosomes
Dormant buds	Trees that grow in temperate climate	These buds enter dormancy and are suitable to freezing. Require little time and labor. Tissue culture is not required	Resistance to cold depends on the season and genotypes. Requires more storage space than other methods. Must have expertise in grafting or budding after storage
In vitro shoot tips	Many types	Can collect tip tissue all year round. Anatomy and physiology requirements are easy to handle	Not developed for all plants. Requires laboratories and skilled workers
Embryonic cultures	Many groups	In general, calli are easy to preserve	Some plants do not produce somatic embryos. Not widely used yet for preserving plant species
Embryonic axis	Plants that are not resistant to desiccation and those with large seeds (recalcitrant seeds)	Easy to excise, remove and process	Tissue culture after storage is required
Zygotic embryos	Plants that do not stand desiccation and are large (recalcitrant seed types)	Removing the seed coat may increase survival	This technique is time consuming and challenging

4.2 Slow Cooling

This method preserves plant varieties in supercool conditions. In nature, plants withstand fall and winter temperatures below freezing, then grow new branches and leaves in spring. As the weather cools, ice crystals form in the water solution between cells inside them. As a result, the remaining solution does not form ice crystals and will have a high concentration of solutes. Therefore, to maintain the water balance inside and outside the cells, water from inside will move out from the protoplasm according to the water potential. With more water outside than inside, cells will dehydrate and wither as their internal solution concentration rises, and the temperature gradually decreases. By the end, unfrozen cells will hold minimal water, while ice crystals outside the membrane will cool to –40°C, suitable for storage in liquid nitrogen (–196°C). The cell solution becomes highly viscous, forming glass-like structures (vitrification) instead of ice crystals (Fig. 4.1).

Thus, preserving plant varieties in a freezing state mimics the principle of slow cooling. This method is used by farmers to culture dormant buds of trees in cold regions by collecting plant materials from trees during winter through chilly weather. The slow cooling method involves exposing the plant material to very low temperatures, such as 0°C, for a short time, or direct immersion in cryoprotectants, such as glycerol and sucrose, at low concentration for 10–20 minutes, then using a programmable freezing controller (Fig. 4.2) to lower the temperature at a rate of 0.5–2°C per minute, depending on crop type and growth stage until –40°C. Then, plant materials will be stored in liquid nitrogen. This method has been applied to non-differentiated tissues, such as cell suspensions and calli. After removing plant materials from liquid nitrogen, the temperature must slowly increase. Since the solution inside the cell still has low water content, if the temperature rises quickly, it may impair the metabolism of the cells.

4.3 Vitrification

The glass transition is a process of physical changes from liquid to non-crystalline, amorphous, and a glass state (Fahy *et al.*, 1984). Cells can be preserved in non-ice crystals, which do not damage the cell membrane (Benson, 2008) (Fig. 4.3). Changing the state of the cell to a glass state

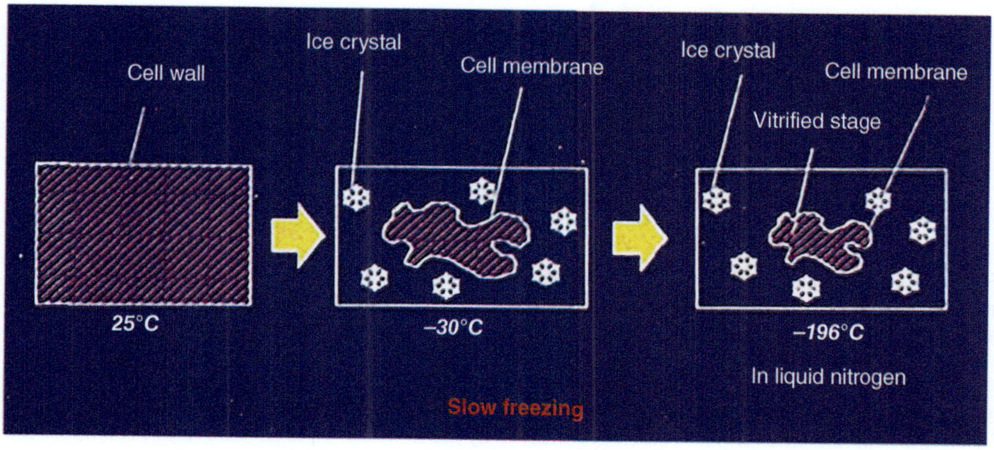

Fig. 4.1. The water inside the cell drops to a point where the solution inside the cell is very concentrated when exposed to extremely low temperatures, such as below 0°C, the water inside the cells changes from liquid to glass-like, which is called the non-crystalline or amorphous phase, also called the vitrification state, and the water molecules will form solid ice crystals. As a result, the amount of water in the solution outside the cell decreases, so the water potential outside the cell is lower than inside the cell. Water then moves from the inside of the cells to outside, enough to prevent the formation of ice crystals inside the cells. The temperature reduction rate must be appropriate to avoid injuring the compartments inside the cell as well as the cell membrane. The cell is still alive. Author's own image.

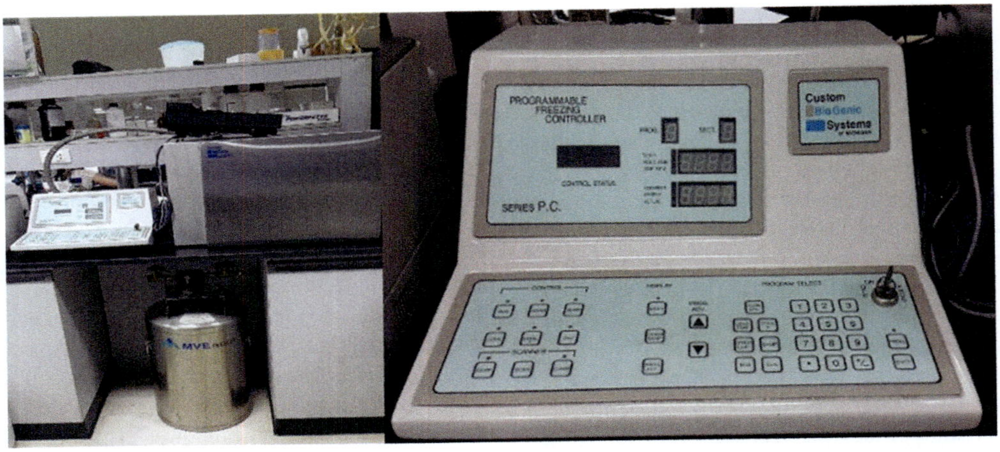

Fig. 4.2. A programmable freezing controller system used for the storage of plant tissues. Author's own image.

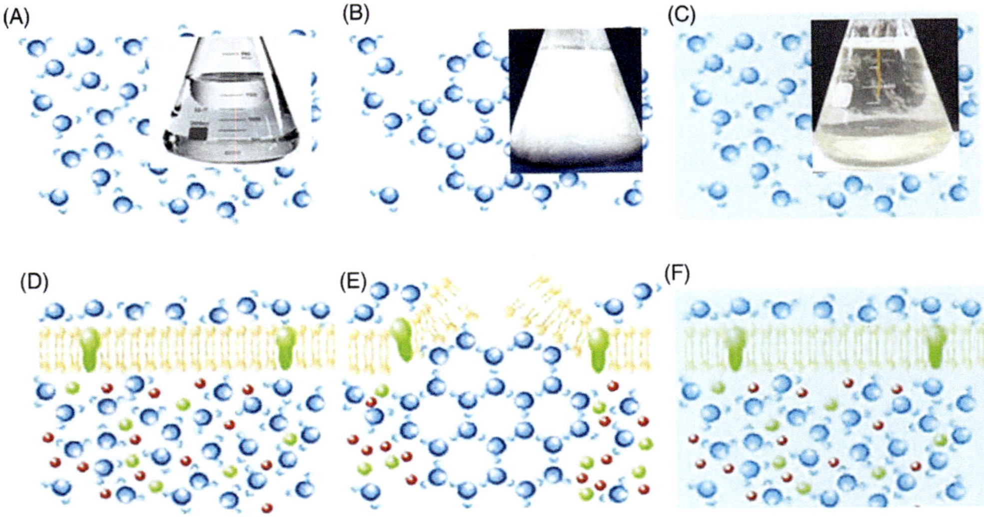

Fig. 4.3. Characteristics and structure of water molecules in a liquid state (A, D), an ice crystal (B, E), and a glass state (C, F). Cell membranes are destroyed when the water inside the cells is in an ice crystal state (E). Author's own image.

requires two conditions: (i) the temperature reduction is sudden, and (ii) the solution within the cell is highly concentrated (Sakai, 2000).

When plant materials in a cryotube are placed into liquid nitrogen, the temperature decreases rapidly (approximately 60°C per second). It is possible to decrease the temperature even quicker using different methods. For instance, the droplet method involves dripping an anti-coolant solution containing the plant materials onto aluminum foil, reducing the temperature by about 4000°C per minute (Panis *et al.*, 2005). Additionally, the cryo-plate method can reduce the temperature by approximately 5000°C per minute (Yamamoto *et al.*, 2011). Moreover, lowering and raising the temperature quickly after removing from liquid nitrogen improves cell survival rates.

The vitrification method, developed by Sakai *et al.* (1990) in Japan, is used for plants from

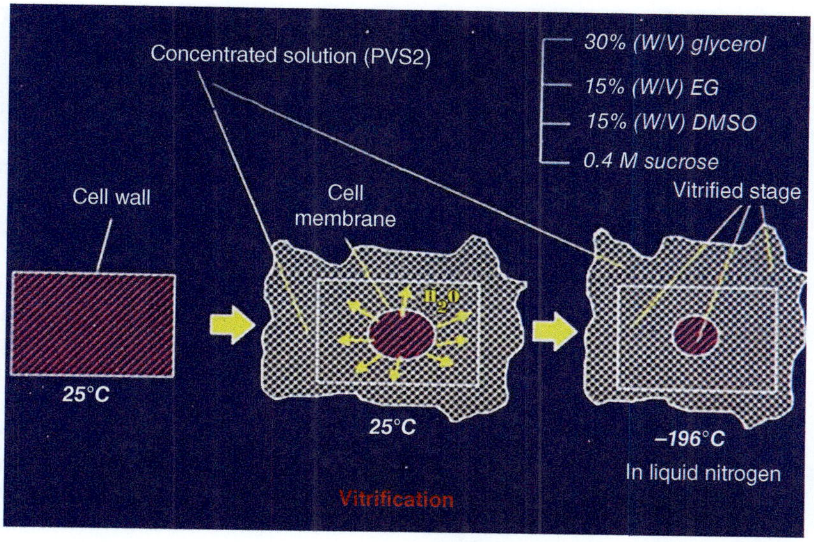

Fig. 4.4. The vitrification method uses concentrated chemicals to prevent cold damage. The most widely used solution is PVS2, developed by Prof. Dr Akira Sakai, containing 30% (w/v) glycerol, 15% (w/v) ethylene glycol (EG), 15% (w/v) dimethyl sulfoxide (DMSO), and 0.4 M sucrose. Glycerol and DMSO will move into the cells. Ethylene glycol will not move into the cells but will cause the water with a higher potential to move out of the cells. Therefore, preserved plant materials in liquid nitrogen are in a glass state inside and outside the cells. Author's own image.

tropical, subtropical, and temperate regions (Fig. 4.4). This process involves immersing plant tissue in a concentrated anti-freezing solution like PVS2 at 25°C and 0°C to dehydrate cells without damage (Matsumoto *et al.*, 1994). The cells moisture content must be around 20–30% (Matsumoto *et al.*, 1995) before storage in liquid nitrogen, forming a glass-like state inside and outside the cells. The vitrification method is easy, convenient, less time-consuming, low-cost, and requires no expensive or complicated equipment (such as a programmable freezing controller). After vitrification, plant materials are removed from the liquid nitrogen and rapidly warmed to prevent the transition from a glass state to ice crystals, thus skipping the liquid state entirely. This process preserves the cell condition; and cryopreserved cells remain viable (Fig. 4.5).

4.4 Steps in Plant Cryopreservation

There are seven significant steps in plant cryopreservation. Some plants may not require every step (skipping steps 1, 2, 5, or 6) in preservation, depending on species and cryopreservation methods. The whole protocol is as follows:

1. Before culturing, the cells will be adjusted at a low temperature by adding 0.3 M sucrose (1–3 days) and then ABA (abscisic acid) to stabilize the cell membrane under dry conditions.

2. Loading solution: Chemicals are used to protect osmotic pressure, such as 2 M glycerol + 0.4 M sucrose (20 minutes) for dehydrating the cells to reduce osmotic pressure, help withstand highly concentrated solutions, such as PVS2, and adjust to dry and vitrification conditions.

3. Dehydration: Scientists now use plant vitrification solutions (PVS) at specific concentrations for each plant material. Common formulas include PVS1, PVS2, PVS3, and PVS4. Solutions like PVS2 are used at 25°C or 0°C to adequately dehydrate cells. These chemicals serve various functions: sugar stabilizes the cell membrane and proteins, dimethyl sulfoxide (DMSO) captures free radicals and protects cell structure, and glycerol stabilizes the cell membrane. However, their mechanisms are not fully understood.

4. Storage in liquid nitrogen: all the cryopreserved cells will be vitrified from −115°C to −196°C (in liquid nitrogen).

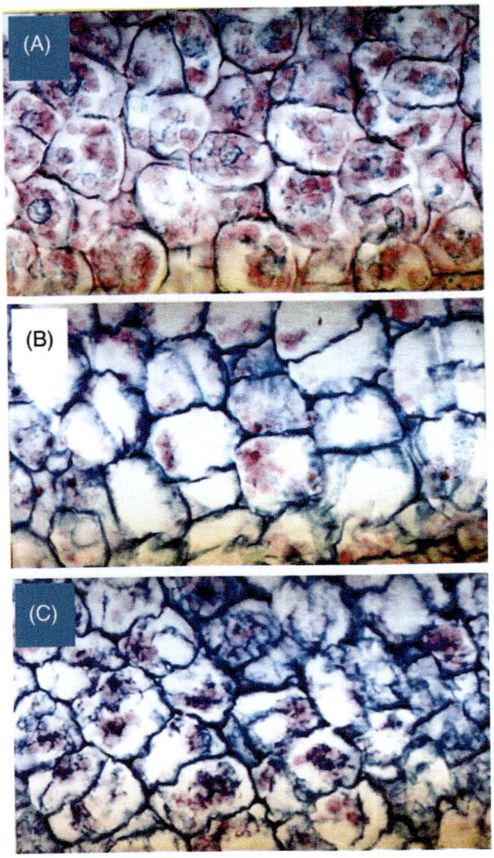

Fig. 4.5. Cryopreservation of jackfruit embryonic tissue using the vitrification method. A) Before storage, cells grow normally. B) Intracellular ice crystals formed after storage in liquid nitrogen without exposure to PVS2 solution, cells die. C) Cells after exposure to PVS2 solution before liquid nitrogen storage. Adapted with permission by Ingenta/UK from Thammasiri, 1999.

5. Rapid warming: warm the cryopreserved cells rapidly at 35–40°C in a water bath to change from a glass to a liquid state.
6. Unloading solution: move the cells to concentrated 1.2 M sucrose for 20 minutes to reduce the osmotic pressure and prevent water from moving into the cells too quickly. This step is for acclimation before culture.
7. Culture the cells on a suitable medium for plantlet development. Then, transfer the plantlets to the greenhouse, applying acclimatization for 2 weeks.

4.5 Methods of Plant Cryopreservation

To preserve plant materials in freezing conditions, it is important to understand the basic mechanisms involved. This process can proceed in two ways: (i) slow cooling or slow freezing (cell dehydration due to cold temperature), and (ii) vitrification (glass formation). In the slow cooling process, dormant buds and slow freezing methods are utilized. In contrast during vitrification, cells are dehydrated by exposure to highly concentrated chemicals or air drying before quick freezing, followed by rapid warming to avoid intracellular ice formation. Six different vitrification-based methods are: (i) vitrification, (ii) encapsulation-dehydration, (iii) encapsulation-vitrification, (iv) droplet-vitrification, (v) vitrification cryo-plate method (V cryo-plate), and (vi) dehydration cryo-plate (D cryo-plate). We summarize these methods below.

4.5.1 The dormant bud method

This is the plant cryopreservation method in which dormant buds are collected during winter when buds were dehydrated at −10 to −30°C before storage in liquid nitrogen (−196 °C). This method was successful in willow (*Salix koriyanagi*) and poplar (*Populus sieboldii*) (Sakai, 1960). Later, the dormant bud method was studied in apple (Sakai and Nishiyama, 1978; Forsline *et al.*, 1998; Towill and Bonnart, 2003), cherry (Towill and Forsline, 1999), and blueberry (Jenderek and Reed, 2017).

The protocol for preserving dormant buds is shown in Fig. 4.6A. This method works well with cold perennials, such as apple, willow, mulberry, and conifers, and it is a good option for plant cryopreservation.

4.5.2 The slow freezing method (revisited)

Slow freezing was the standard method in the early days, needing a programmable controller to reduce temperature at 0.5–2°C per minute until about −40°C, then storing in liquid nitrogen. This lengthy process required expensive tools and

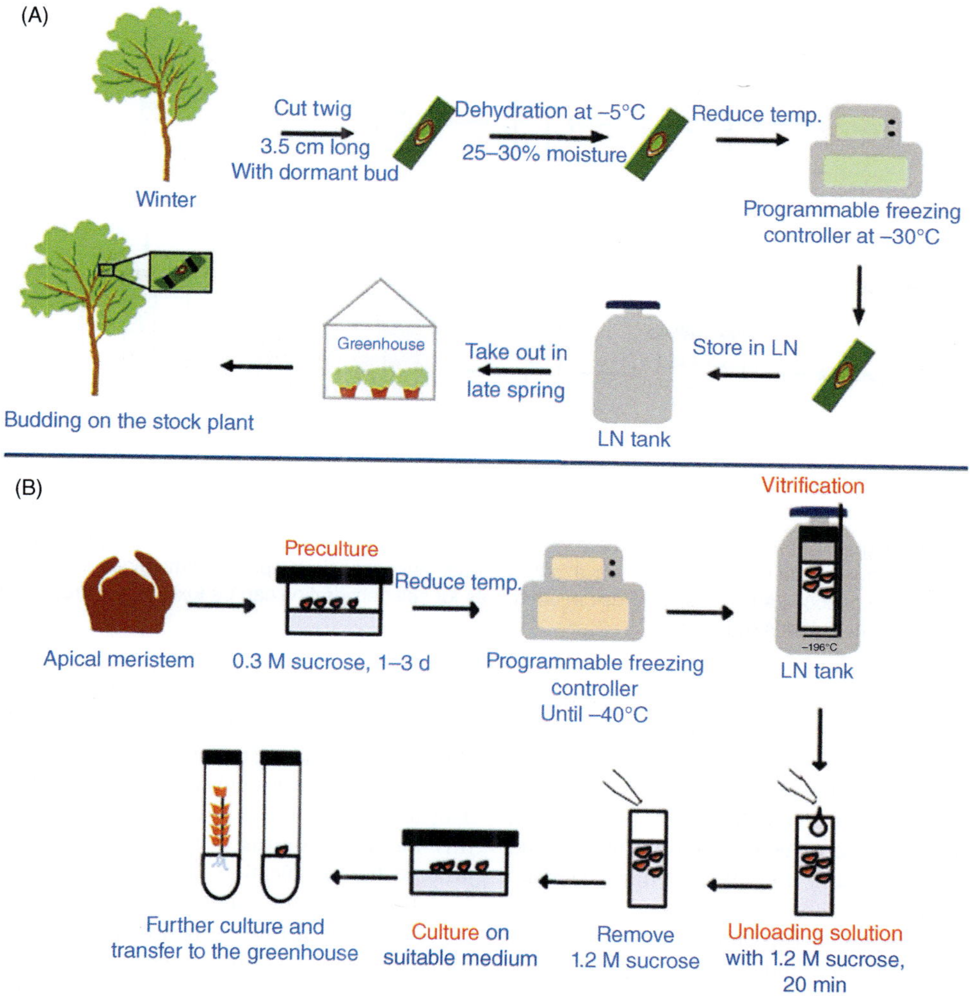

Fig. 4.6. A) Diagram of dormant bud cryopreservation method. B) Diagram of the slow freezing method. The dormant bud method takes advantage of physiological changes within the plant that allow adaptation to the winter, such as reduced moisture content and tissue dormancy. These changes facilitate ensuing freezing and storage in liquid nitrogen (LN). The slow freezing method mimics the natural process of dormancy by reducing the moisture culture of apical meristems with an osmotic protectant such as sucrose, followed by freezing a programmable controller. The effectiveness and reliability of this latter method are extremely high.

was popular from 1980 to 1990 for preserving undeveloped plant tissues like suspended cells and calli. Fig. 4.6B illustrates the slow freezing method, which preserves plant species in icy conditions. Although it is time consuming and requires a programmable freezing controller, this method can be effective. In Beijing's National Gene Bank, slow freezing of mulberry tissues achieves a 90% survival rate (Zhang *et al.*, 2014).

4.5.3 The vitrification method (revisited)

Vitrification is a method developed to lower the temperature and then warm rapidly, ensuring that all cells and tissues enter a glass state (Fig. 4.7). The vitrification method involves treating explants with plant vitrification solutions, such as PVS1, PVS2, PVS3, or PVS4, to induce dehydration during cooling and warming,

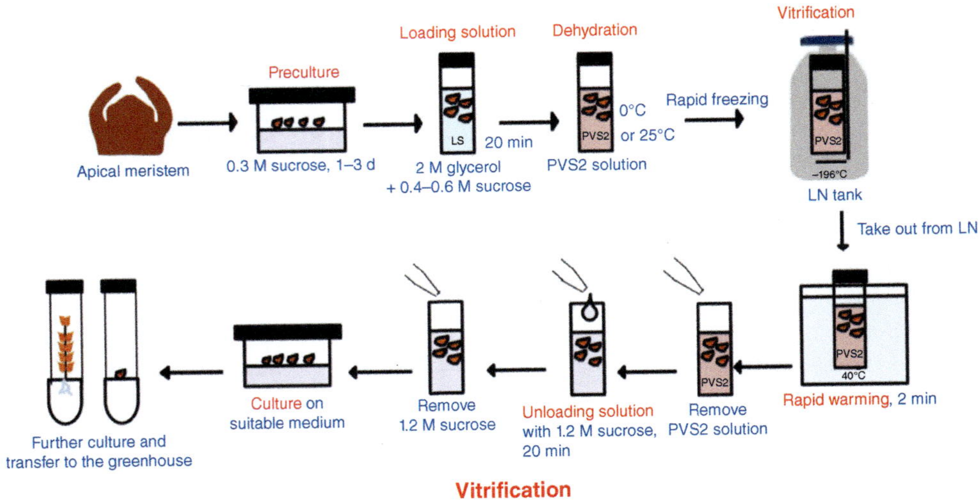

Fig. 4.7. Diagram of the vitrification method. Vitrification is designed to have a short processing time, low cost, and requires little work. It results in high survival rates but necessitates precise timing due to the brevity of the protocol. Additionally, the use of chemicals for cell dehydration can be toxic to cryopreserved plant materials. Author's own image.

thereby preventing intracellular ice-crystal formation (Usagami *et al.*, 1989; Sakai *et al.*, 1990). Successful cryopreservation using the vitrification method relies on optimizing exposure time to PVS for effective dehydration (Niino *et al.*, 2007) because excessive exposure to PVS can lead to cell injury and intracellular ice formation during cooling.

The optimum exposure time to PVS depends on explant size and is species-specific. The suitable dehydration duration was related to the sample size, the composition, and loading solution (Chen and Wang, 2002). Sakai *et al.* (1990) succeeded in cryopreserving nucellar cells of naval oranges using the PVS2 solution; after that, many plants were experimented with successfully.

4.5.4 The encapsulation-dehydration method

The encapsulation-dehydration method was developed for artificial seed production, in which explants are encapsulated in alginate beads, precultured with high sucrose, desiccated by air-drying in a laminar air-flow cabinet or with silica gel, and then plunged into liquid nitrogen (Sakai *et al.*, 2000; Matsumoto and Sakai, 1995) (Fig. 4.8). This method's advantages are easy manipulation of encapsulated explants (Hirai *et al.*, 1998), preventing direct contact of toxic chemicals with plant materials, and non-toxic cryoprotectants are applied to protect during dehydration (Niino and Sakai, 1992). However, this method requires a longer dehydration process than vitrification (Thammasiri, 2000). Nonetheless, Fabre and Dereuddre (1990) successfully used this method to cryopreserve tomato apical meristems, suggesting that this method could be applied to many other plant species.

More recently, Jitsopakul *et al.* (2008a) successfully cryopreserved protocorms of the popular orchid *Vanda coerulea* by the encapsulation-dehydration method in combination with a loading solution (Fig. 4.9). Protocorms were selected 70 days after sowing the seeds, which were harvested from 7-month-old capsules. After encapsulation in an alginate matrix composed of 2% Na-alginate, 2 M glycerol plus 0.4 M sucrose (the loading solution), the protocorms were precultured in modified VW liquid medium (Vacin and Went, 1949) supplemented with 0.7 M sucrose on a shaker (100 rpm) at 25 ± 3°C for 20 h. Encapsulated protocorms were then dehydrated in a laminar flow cabinet at 25 ± 3°C for 1–10 h, then directly plunged into liquid nitrogen for 1 day. After thawing at 40°C for

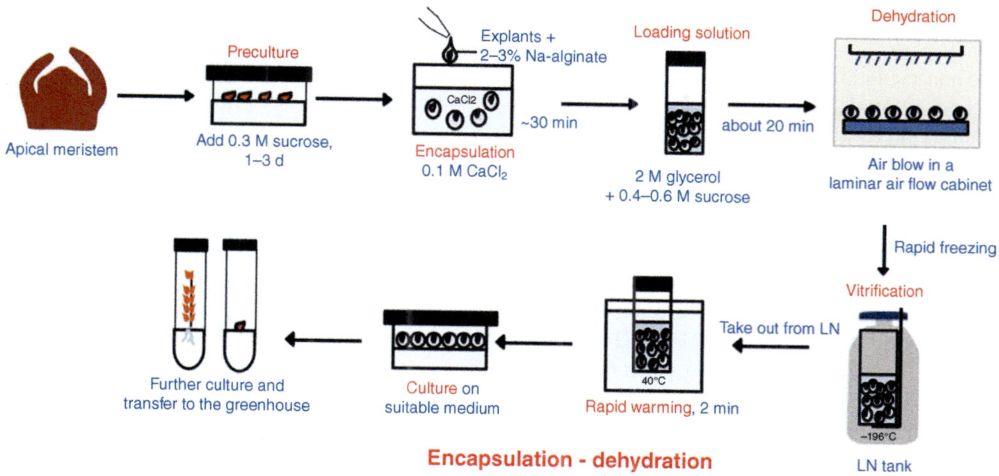

Fig. 4.8. Diagram of the encapsulation-dehydration method. The explants are encapsulated in an alginate bead and then sequentially dehydrated by air flow and exposure to highly concentrated PVS. Eventually explants are warmed, cultured, and allowed to regenerate. Author's own image.

Fig. 4.9. Regrowth of *Vanda coerulea* protocorms after cryopreservation by encapsulation-dehydration in combination with loading solution. A) Precultured beads after sterile air-flow dehydration for 10 h. B) Cryopreserved protocorms after 20 days of regrowth. C) 3 months of culture on modified VW agar medium showing shoot growth. D) Plantlets after 8 months of culture on modified VW agar medium (left: non-cryopreserved, right: cryopreserved plantlet). E) Plantlets derived from cryopreserved protocorms after 5 months. F) 15 months of culture in the greenhouse. Bars: A–C = 1 mm, D = 0.5 cm, E and F = 1 cm. Adapted with permission by Ingenta/UK from Jitsopakul *et al.*, 2008a.

2 min, cryopreserved beads were cultured on modified VW agar medium for regrowth. The highest regrowth rate (40%) was observed with cryopreserved beads with a 35% moisture that were subject to 8 h of dehydration. No morphological variation was detected between non-cryopreserved and cryopreserved plantlets, and the ploidy level was not altered either.

4.5.5 Encapsulation-vitrification method

The encapsulation-vitrification method combines encapsulation-dehydration and vitrification. The explants are encapsulated in an alginate bead and then subjected to dehydration by highly concentrated plant vitrification solutions, such as PVS1, PVS2, PVS3, and PVS4. This method shows a higher survival rate than encapsulation-dehydration (Sakai *et al.*, 2008; Matsumoto, 2017).

In this method chemicals are not used for dehydration, but an air flow is used instead; therefore, it is not toxic to either plant materials or people. Encapsulation helps to reduce the damage from handling plant materials, while air flow dehydration takes 30 minutes, which works as a buffer for keeping the process flexible. In addition, there is no unloading solution after rapid warming since the water from outside the cells will not move in quickly due to encapsulation with calcium alginate and air flow dehydration. The disadvantages for this method are the long operation for 2–3 days, a medium to low survival depending on how tightly the protocols are adjusted, and a high sucrose concentration used after encapsulation, which several plant species do not tolerate.

A diagram for the encapsulation-vitrification method is shown in Fig. 4.10. The encapsulation-vitrification method is a lengthy process of chemical encapsulation, dehydration and culture. Survival is comparatively higher than in the encapsulation-dehydration method and close to the regular vitrification method.

4.5.6 The droplet-vitrification method

This recent method involves sudden freezing of small drops of PVS on aluminum foil. It also combines droplet-freezing and solution-based vitrification. The sample is placed in a droplet of vitrification solution held by an aluminum foil strip, and then it is frozen by rapid immersion in liquid nitrogen (Sakai and Englemann, 2007). Then, rapid warming is done by dipping the aluminum foil strips in the unloading solution without using a water bath (Kim *et al.*, 2006), which prevents recrystallization of intracellular ice or additional cell dehydration by extracellular ice (Kim *et al.*, 2006). The rationale for this protocol is that fast heat transfer is needed during the cooling and warming procedures to avoid freezing injury (Agrawal *et al.*, 2004; Kim *et al.*, 2006), and quite conveniently, aluminum foil has a high thermal conductivity, resulting in quick and uniform heat distribution among tissue (Halmagyi *et al.*, 2005; Kim *et al.*, 2006;). The first instance of success by the droplet-vitrification method was reported in potatoes (Pennycooke and Towill, 2000). Later on, reports

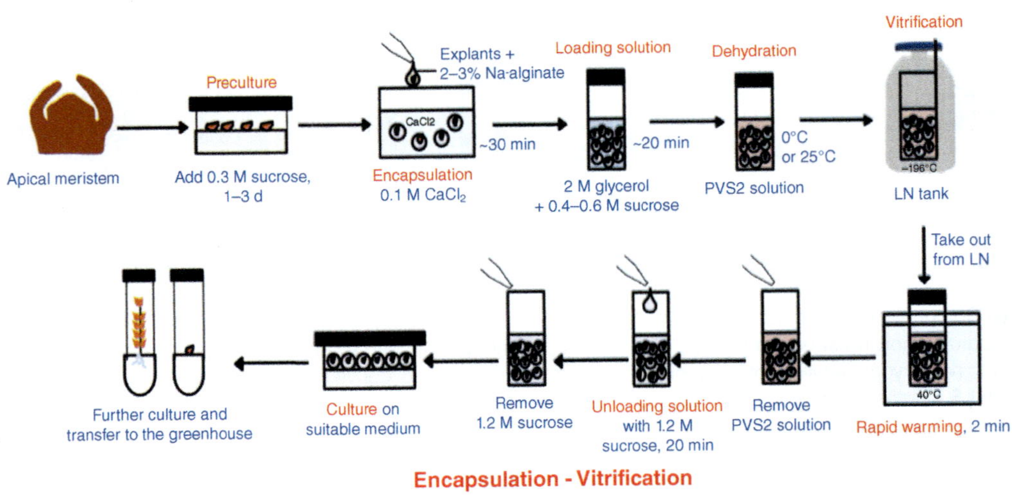

Fig. 4.10. Diagram of the encapsulation-vitrification method. This method involves sequential encapsulation, loading into glycerol and sucrose, chemical dehydration, vitrification, warming, exposure to sucrose, culture, and regeneration.

emerged in papaya (Ashmore *et al.*, 2001), prunes (De Boucaud *et al.*, 2002), jam (Leunufna and Keller, 2003), banana (Panis *et al.*, 2005), sugar cane (Barraco *et al.*, 2011), oil palm (Gantait *et al.*, 2015), grapes (Pathirana *et al.*, 2015), and oranges (Volk *et al.*, 2012), and then ornamentals such as chrysanthemum (Halmagyi *et al.*, 2004), roses (Halmagyi and Pinker, 2006), and of course valuable orchids such as the Chinese ground orchid *Bletilla striata* (Jitsopakul *et al.*, 2008b), the Tiger orchid *Grammatophyllum speciosum* (Sopalun *et al.*, 2010), the blue orchid *Vanda coerulea* (Jitsopakul *et al.*, 2008a), and vanilla (Gonzalez-Arnao *et al.*, 2009; Hernandez-Ramirez *et al.*, 2014).

In Thailand, Jitsopakul *et al.* (2008b) studied the droplet-vitrification method for cryopreservation of orchid *Bletilla striata* mature seeds (0 day after sowing), zygotic embryos (3 days after sowing), and protocorms (6, 9, and 12 days after sowing) (Fig. 4.11). Mature seeds were surface-sterilized and sown on solidified New Dogashima (ND) medium (Tokuhara and Mii, 1993) supplemented with 3% sucrose and cultured under a photosynthetic photon flux (PPF) of 62 μmol m^{-2} s^{-1}, for 16 h per day, and at 25°C. After that mature seeds, zygotic embryos, and 6-day-old protocorms were sequentially precultured in liquid ND medium supplemented with 0.3 M sucrose for three hours on a rotary shaker (110 rpm), and then dehydrated with 2 M glycerol and 0.4 M sucrose in ND liquid medium (as a loading solution) for 15 min, followed by exposure to the PVS2 solution for 60 min at 25°C. Finally, plant materials were soaked in liquid nitrogen using the droplet-vitrification method and cultured on solidified ND medium supplemented with 3% sucrose. The survival rate of cryopreserved mature seeds, zygotic embryos, and 6-day-old protocorms were 93%, 91%, and 84%, respectively. The survival rate of older protocorms was low or close to zero. Figure 4.11A–E shows the normal development program of mature seeds in *Bletilla* sp. Usually mature seeds develop into zygotic embryos 3 days after sowing (Fig. 4.11B). Then protocorms form from green spots on zygotic embryos at 6 days of sowing (Fig. 11C). Protocorms develop apical meristems at 9 days of sowing (Fig. 4.11D) and form primary leaves at 12 days of sowing (Fig. 4.11E).

A diagram for the droplet-vitrification method is shown in Fig. 4.12. The droplet-vitrification method is a new method that was developed from the old vitrification protocol. The explants

Fig. 4.11. Mature seeds of *Bletilla striata*. A) Development of mature seeds sown on solidified ND medium supplemented with 3% sucrose under illumination provided at an intensity of 62 μM m^{-2} s^{-1} for 16 h per day at 25°C for 3 days. B) 3 days. C) 6 days. D) 9 days. E) 12 days. F) Twenty droplets of PVS2 solution (2 μl) with protocorms, placed on sterilized aluminum foil strip (7 x 20 mm^2). Bars: A–E = 0.5 mm, F = 1 cm. Figure adapted with permission by Ingenta/UK, from Jitsopakul *et al.*, 2008b.

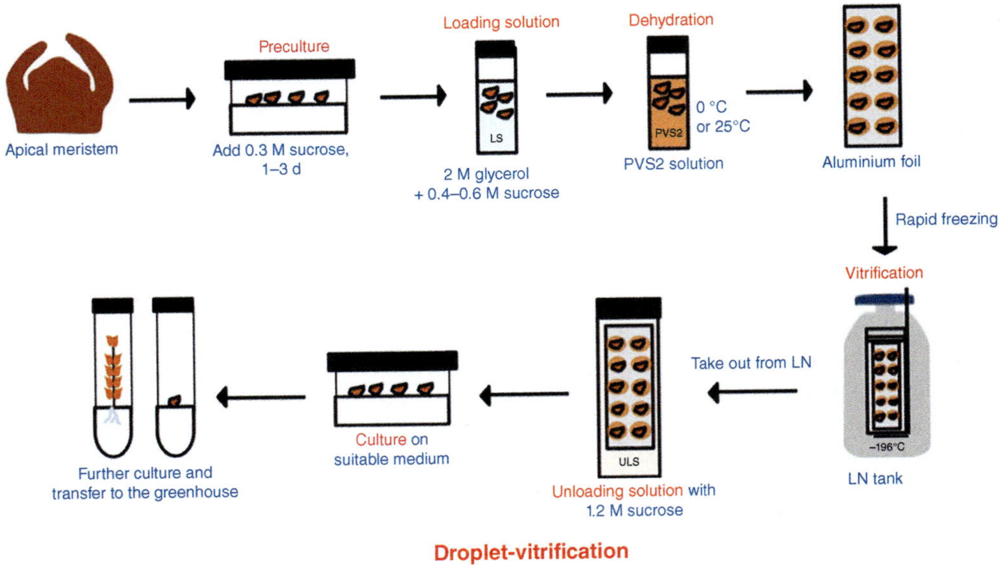

Fig. 4.12. Diagram of the droplet-vitrification method. The explants are dehydrated with one drop of concentrated plant vitrification solution and placed on aluminum foil for heat transfer. This method ensures a high survival rate. Author's own image.

are dehydrated using just one drop of highly concentrated plant vitrification solution and placed on an aluminum foil that contributes to heat transfer. Moving about 10–12 explants at once onto the aluminum foil is convenient. High survival is obtained from the rapid cooling and warming achieved by the aluminum foil, consequently a rapid warming step is not necessary with this protocol.

4.5.7 The V cryo-plate method

An innovative approach for preserving plant species by freezing was developed by Yamamoto *et al.* (2011), which is usually referred to as the V cryo-plate method (Fig. 4.13A). It combines into one the procedures of encapsulation-vitrification and droplet-vitrification to facilitate the adhesion of artificial seeds to aluminum sheets (the so called cryo-plate). The plate has the following dimensions: 7 mm (width) × 37 mm (length) × 0.5 mm (height), with 10–12 holes with a size of 1.5 mm × 0.75 mm. These sheets are thick to improve the thermal conductivity rate, and as a result the cooling rate is approximately 5000°C per min while the warming rate is close to 4500°C per min. Therefore, the V cryo-plate method leads to higher survival rates than the droplet-vitrification method. Moreover, the protocol is easy and convenient because the plant materials are attached onto the aluminum sheet throughout the operation and not suspended in solution such as PVS and PVS3. This method does not require specific skills either.

4.5.8 The D cryo-plate method

Niino *et al.* (2013) developed the D cryo-plate method (Fig. 4.13B), which combines encapsulation-dehydration with the V cryo-plate method (Niino *et al.*, 2014). Occasionally there is low survival after the V-cryo plate method as some plant species may be affected by using the PVS2 solution. Therefore, to avoid this damage, cells are dried by blowing air from a laminar air-flow cabinet instead. In addition, the cryopreserved plant materials can be larger than those preserved with the V cryo-plate method because there are not any problems related to toxicity to the PVS2 solution after prolonged immersion. Thus, the V and D cryo-plate methods follow the same protocol except for the cell dehydration step by air-drying.

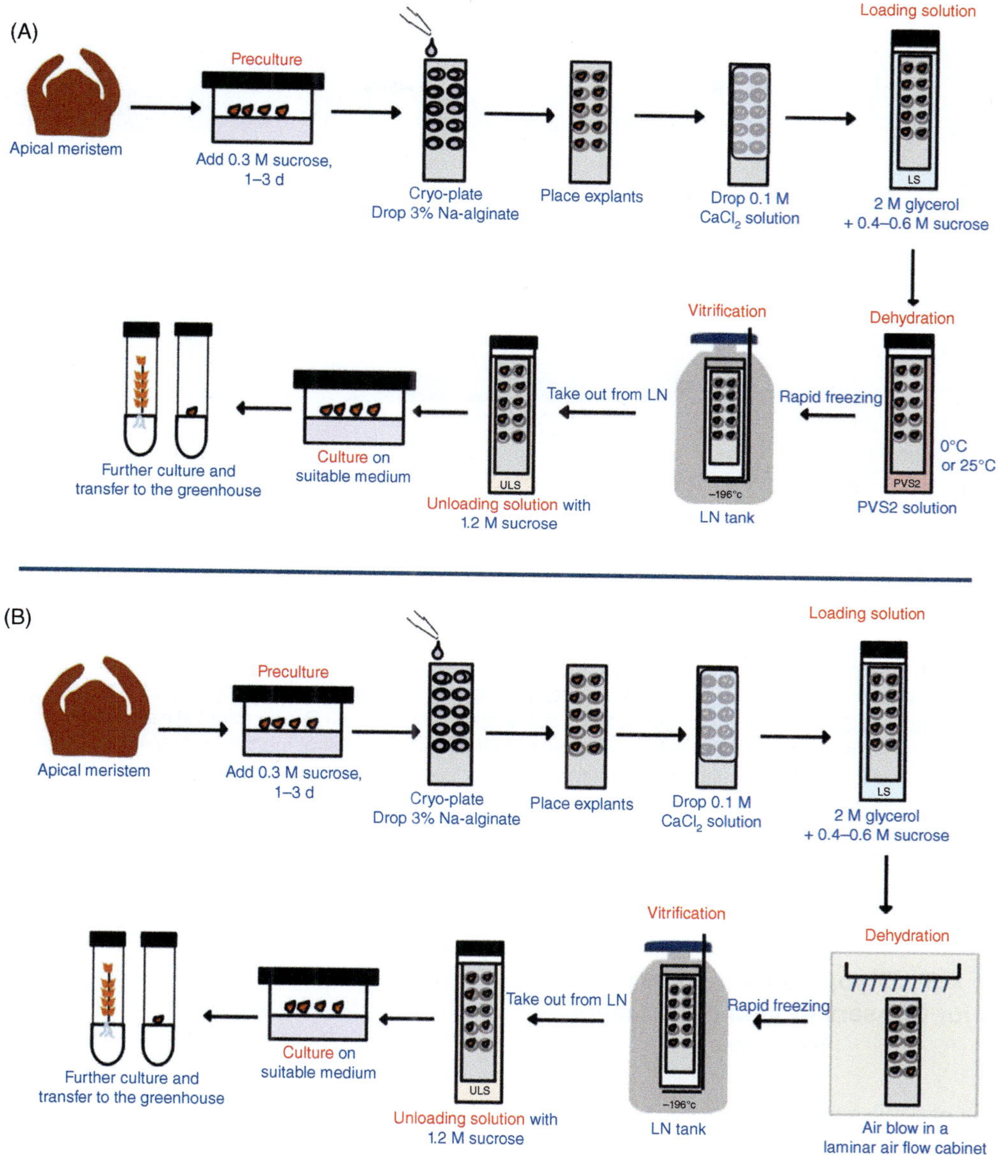

Fig. 4.13. Diagram of the V and D cryo-plate method. A) the V-cryo plate, this method is an iteration of the encapsulation and droplet-vitrification protocols. It features loading on osmo-protectants such as glycerol and sucrose and requires the use of only one super coolant solution, the PVS2 solution. This method leads to high survival rates. B) the D cryo-plate method, this is a combination of the encapsulation-dehydration with V cryo-plate methods. It uses a thick aluminum sheet plate, which conducts heat extremely well. The result is a further increase in survival rates. Author's own image.

Later, Cordova and Thammasiri (2016) developed the D cryo-plate method using silica gel and drying beads to dehydrate cells of the orchid *Arundina graminifolia* (Fig. 4.14). Regrowth was observed in the second week of transfer to half-strength MS media. Regrowth was 73.8% and 76.5% using silica gel and drying beads for dehydration, respectively. Dehydration using silica gel or drying beads did not significantly affect the regrowth rate. Protocorms dehydrated using silica gel or drying beads developed into normal plantlets.

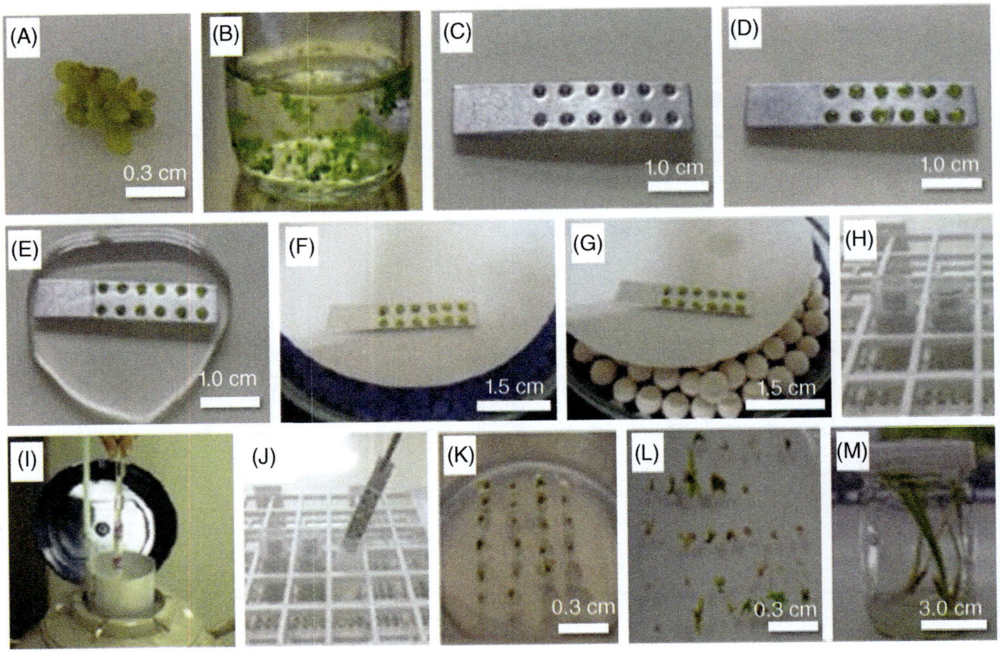

Fig. 4.14. Cryopreservation of orchid *Arundina graminifolia* by the cryo-plate method but with silica gel or drying beads for dehydration. A) Protocorm development. B) Preculture of protocorms in half-strength Murashige and Skoog (MS) liquid medium with 0.7 M sucrose for 1 d. C) Pouring the alginate solution containing 2% (w/v) sodium alginate in calcium-free half-strength MS basal medium with 0.4 M sucrose in the wells. D) Placing the precultured protocorms in the wells one by one. E) Pouring the calcium chloride solution containing 0.1 M calcium chloride in half-strength MS basal medium with 0.4 M sucrose. F) Dehydration with 50 g silica gel. G) Dehydration with 30 g drying beads. H) Placing every cryo-plate inside a 2 ml cryotube. I) Plunging 2 ml cryotubes into liquid nitrogen for 1 d. J) Warming in 1.2 M sucrose solution for 20 minutes. K) Plating on half-strength MS agar medium. L) Regrowth. M) Regrowth after 60 days. Adapted with permission by Ingenta/UK, from Cordova and Thammasiri (2016).

4.6 Cryopreservation as a Tool for Conservation of Plant Genetic Resources

The conservation of ornamental plant genetic resources is of paramount importance due to the significant decline in their populations within natural habitats, caused by natural disasters, deforestation, and the extraction of plants for commercial purposes. Currently, ornamental plant germplasms are preserved through *in situ* (national parks and conservation areas), and *ex situ* methods, supported by government organizations, academic institutions, the private sector, and devoted individuals. *Ex situ* conservation strategies include running living collections, seed banks, and *in vitro* conservation via tissue culture and cryopreservation. My colleagues and I have conducted studies and reported on the cryopreservation of pollen (in species *Amorphophallus koratensis*), pollinia, seeds, protocorms, protocorm-like bodies, and shoot tips of various Thai orchid species and hybrids (Table 4.2). Many of the orchid genera and species in which cryopreservation has been evaluated are of high commercial value, such as the Pigeon orchid *Dendrobium crumenatum* Sw., the Deer-antlered orchid *Phalaenopsis cornu-cervi* (Breda) Blume & Rchb. F, the Foxtail orchid *Rhynchostylis gigantea* (L.) Ridl., and *Vanda tricolor* Lindl. thus this information may have great commercial application and impact.

4.7 Conclusions

The reproduction of elite ornamentals and orchids necessitates consistent clonal reproduction

Table 4.2. Successful cryopreservation of ornamental plant *Amorphophallus koratensis* and selected orchid species and hybrids from Thailand. Author's own table.

Species	Materials	Method	Dehydration	Survival/regrowth	References
Aerides houlettiana Rchb.f.	Pollinia	V cryo-plate	60 min PVS2	100% survival, 67% capsule set	Jitsopakul *et al.* (2021)
Aerides koratensis	Pollen	Direct plunged into liquid nitrogen, 7 d	Collect pollen at the first day after anthesis	34% pollen viability	Soonthornkalump *et al.* (2020)
Arundina graminifolia	Protocorms	D cryo-plate	30 g of drying beads in the laminar air-flow cabinet	77% regrowth	Cordova and Thammasiri, (2016).
Cymbidium finlaysonianum Wall. ex Lindl.	Protocorm-like bodies (PLBs)	Encapsulation	Stored at 8 ± 2°C, for 105 d	44% conversion frequency	Klaocheed *et al.* (2018)
Cymbidium finlaysonianum Wall. ex Lindl.	PLBs	Vitrification	Precultured with 0.5 M sucrose for 2 days, 60 min on PVS2 at 0°C	33.5% regrowth	Rittirat *et al.* (2019)
Coelogyne trinervis L.	Pollinia	V cryo-plate	20 min on PVS2	100% survival, 100% capsule set	Jitsopakul *et al.* (2021)
Dendrobium cariniferum	Protocorms	Encapsulation-vitrification	60 min on PVS2 at 25 ± 2°C	15% survival	Pornchuti and Thammasiri, (2008)
Dendrobium chrysotoxum L.	Pollinia	V cryo-plate	20 min PVS2	40% survival, 20% capsule set	Jitsopakul *et al.* (2021)
Dendrobium crumenatum Sw.	PLBs	Encapsulation	Stored at 8 ± 2°C, for 105 d	50% conversion frequency	Klaocheed and Rittirat (2018)
Dendrobium draconis Rchb. f.	Protocorms	V cryo-plate	Preculture with 0.4 M sucrose for 3 d, 0.6 M in LS, 10 min in PVS2	75% survival	Jitsopakul *et al.* (2020)
Dendrobium cruentum Rchb. f.	Seeds	D cryo-plate	60 min with drying beads inside the laminar air-flow cabinet	69% survival, 58% germination	Prasongsom *et al.* (2020)
Dendrobium signatum Rchb. f.	Pollinia	V cryo-plate D cryo-plate	40 min on PVS2 3 h in the laminar air-flow cabinet	55.6% survival 50% survival	Jitsopakul *et al.* (2019)
Dendrobium Walter Oumae	shoot tips	Encapsulation-dehydration	6–8 h in the laminar air-flow cabinet	13% survival	Lurswijidjarus and Thammasiri, (2004)
Eriabractescens L.	pollinia	V cryo-plate	20 min on PVS2	100% survival, 66.7% capsule set	Jitsopakul *et al.* (2021)

Continued

Table 4.2. Continued.

Species	Materials	Method	Dehydration	Survival/regrowth	References
Grammatophyllum speciosum	Protocorms	Encapsulation-vitrification	60 min on PVS2	14% regrowth	Sopalun *et al.* (2010)
G. speciosum	Protocorms	Droplet-vitrification	30 min on PVS2	38% regrowth	Sopalun *et al.* (2010)
G. speciosum	Seeds	D cryo-plate	120 min alongside drying beads of silica gel	87% viability rate	Thanasuttanithi and Thammasiri, (2022)
Paphiopedilum exul (Ridl.) Rolfe	Seeds	Encapsulation-vitrification	40 min on PVS2	29.7% seed germination, growth index (GI) at 2.03	Imsomboon and Thammasiri (2020)
Phalaenopsis cornu-cervi (Breda) Blume & Rchb. f.	PLBs	Encapsulation	Stored at 25 ± 1°C, 30 d	100% germination	Rittirat *et al.* (2015)
Pomatocalpa spicatum Breda.	Pollinia	V cryo-plate	40 min on PVS2	100% survival, 100% capsule set	Jitsopakul *et al.* (2021)
Rhynchostylis gigantea (L.) Ridl.	Pollinia	V cryo-plate	40 min on PVS2	100% survival, 100% capsule set	Jitsopakul *et al.* (2021)
Seidenfadenia mitrata (Rchb.f.) Garay.	pollinia	V cryo-plate	40 min on PVS2	100% survival, 50% capsule set	Jitsopakul *et al.* (2021)
Spathoglottis plicata Blume.	Pollinia	V cryo-plate	20 min on PVS2	100% survival, 100% capsule set	Jitsopakul *et al.* (2021)
Vanda coerulea	Seeds	Vitrification	70 min on PVS2 at 25 ± 2°C	67% germination	Thammasiri and Soamkul (2007)
V. tricolor	Seeds	Vitrification	180 min and PVS2 on ice	13.6% germination	Jitsopakul *et al.* (2012)
V. coerulea	Protocorms	Encapsulation-dehydration	8 h in the laminar air-flow cabinet	40%	Jitsopakul *et al.* (2008a)
V. bensonii Batem.	Pollinia	V cryo-plate	40 min on PVS2	100% survival, 100% capsule set	Jitsopakul *et al.* (2021)
V. brunnea Rchb.f.	Pollinia	V cryo-plate	60 min on PVS2	100% survival, 100% capsule set	Jitsopakul *et al.* (2021)

and cross-pollination. Additionally, certain species are endangered due to deforestation and illegal trade practices. Cryopreservation presents a cost-effective alternative for managing forested areas, with various protocols providing flexibility concerning the species and tissue selection. Furthermore, these methods can be applied to high-demand food crops. Consequently, research into cryopreservation holds broad commercial and scientific applicability. It is anticipated that these findings will prove beneficial to breeders and researchers across Asia and the world.

References

Agrawal, A., Swennen, R. and Panis, B. (2004) A comparison of four methods for cryopreservation of meristems in banana (*Musa* spp.). *CryoLetters* 25, 101–110.

Ashmore, S., Azimi, M. and Drew, R.A. (2001) Cryopreservation trials in *Carica* papaya. *Acta Horticulturae* 560, 117–120.

Barraco, G., Sylvestre, I., and Engelmann, F. (2011) Comparing encapsulation-dehydration and droplet-vitrification for cryopreservation of sugarcane (*Saccharum* spp.) shoot tips. *Scientia Horticulturae* 130, 320–324.

Benson, E.E. (2008) Cryopreservation theory. In: Reed, B.M. (ed.) *Plant Cryopreservation: A Practical Guide*. Springer, New York, pp. 15–32.

Chen, Y. and Wang, J.H. (2002) Cryopreservation of carrot (*Daucus carota* L.) cell suspensions and protoplasts by vitrification. *CryoLetters* 24, 57–64.

Cordova II, L.B. and Thammasiri, K. (2016) Cryopreservation on a cryo-plate of *Arundina graminifolia* protocorms, dehydrated with silica gel and drying beads. *CryoLetters* 37(2), 68–76.

Cronquist, A. (1981) *An Integrated System of Classification of Flowering Plants*. Columbia University Press, New York, USA.

de Boucaud, MT., Brison, M., Helliot, B. and Hervé-Paulus, V. (2002) Cryopreservation of Prunus. In: Towill, L.E. and Bajaj, Y.P.S. (eds) *Cryopreservation of Plant Germplasm II. Biotechnology in Agriculture and Forestry*, vol 50. Springer, Berlin, Germany. DOI: 10.1007/978-3-662-04674-6_21

Engelmann, F. (2004) Plant cryopreservation: progress and prospects. *In Vitro Cellular and Developmental Biology-Plant* 40, 427–433.

Fabre, J. and Dereuddre, J. (1990) Encapsulation-dehydration: a new approach to cryopreservation of *Solanum* shoot-tips. *CryoLetters* 11, 413–426.

Fahy, G.M., MacFarlane, D.R., Angell, C.A. and Meryman, H.T. (1984) Vitrification as an approach to cryopreservation. *Cryobiology* 21, 407–426.

Forsline, P.L., Towill, L.E., Waddell, J.W., Stushnoff, C., Lamboy, W.F. and McFerson, J. R. (1998) Recovery and longevity of cryopreserved dormant apple buds. *Journal of the American Society for Horticultural Science* 123, 365–370.

Gantait, S., Sinniah, U.R., Suranthran, P., Palanyandy, S.R. and Subramariam, S. (2015) Improved cryopreservation of oil palm (*Elaeis guineensis* Jacq.) polyembryoids using droplet vitrification approach and assessment of genetic fidelity. *Protoplasma* 252, 89–101. DOI: 10.1007/s00709-014-0660-x

Gonzalez-Arnao, M.T., Lazaro-Vallejo, C.E., Engelmann, F., Gamez-Pastrana, R., Martinez-Ocampo, Y. and Diaz-Ramo, C. (2009) Multiplication and cryopreservation of vanilla (*Vanilla planifolia* 'Andrews'). *In Vitro Cellular and Developmental Biology-Plant* 45, 574–582. DOI: 10.1007/s11627-009-9242-6

Halmagyi, A. and Pinker, I. (2006) Plant regeneration from *Rosa* shoot tips cryopreserved by a combined droplet-vitrification method. *Plant Cell, Tissue and Organ Culture* 84, 145–153.

Halmagyi, A., Fishen-Klüver, G., Mix-Wagnen, G. and Schumacher, H.M. (2004) Cryopreservation of *Chrysanthemum morifolium* (*Dendranthema grandiflora* Ramat.) using different approaches. *Plant Cell Reports* 22, 371–375.

Halmagyi, A., Deliu, C., and Coste, A. (2005) Plant regrowth from potato shoot tips cryopreserved by a combined vitrification-droplet method. *CryoLetters* 26, 313–322.

Hernández-Ramírez, F., González-Arnao, M.T., Crez-Cruz, C., Pastelín-Solano, M. and Engelmann, F. (2014) Comparison of different preconditioning and loading treatments with vanilla (*Vanilla planifolia* Jack.) apices cryopreserved using the droplet-vitrification procedure. *Acta Horticulturae* 1039, 173–180.

Hirai, D., Shirai, K., Shirai, S. and Sakai, A. (1998) Cryopreservation of *in vitro*-grown meristems of strawberry (*Fragaria* x *ananassa* Duch.) by encapsulation-vitrification. *Euphytica* 101, 109–115.

Imsomboon, T. and Thammasiri, K. (2020) Cryopreservation of *Paphiopedilum exul* (Ridl.) Rolfe seeds using encapsulation-vitrification and encapsulation-dehydration methods. *Acta Horticuturae* 1298, 195–204.

Jenderek, M.M. and Reed, B.M. (2017) Cryopreserved storage of clonal germplasm in the USDA national plant germplasm system. *In Vitro Cellular and Developmental Biology-Plant* 53, 299–308.

Jitsopakul, N., Thammasiri, K., and Ishikawa, K. (2008a) Cryopreservation of *Vanda coerulea* protocorms by encapsulation-dehydration method. *CryoLetters* 29, 253–260.

Jitsopakul, N., Thammasiri, K., and Ishikawa, K. (2008b) Cryopreservation of *Bletilla striata* mature seeds, 3-day germination, and protocorms by droplet-vitrification. *CryoLetters* 29, 517–526.

Jitsopakul, N., Thammasiri, K., Yukawa, T. and Ishikawa, K. (2012) Effect of cryopreservation on seed germination and protocorm development of *Vanda tricolor*. *ScienceAsia* 38, 244–249.

Jitsopakul, N., Sangyojarn, P., Homchan, P. and Thammasiri, K. (2019) Efficiency of aluminum cryo-plates for cryopreservation of *Dendrobium signatum* Rchb. f. pollinia. *Acta Horticulturae* 1234, 279–286.

Jitsopakul, N., Homchan, P. and Thammasiri, K. (2020) Cryopreservation of *Dendrobium draconis* Rchb. f. protocorms using the V cryo-plate method. *Acta Horticulturae* 1298, 213–217.

Jitsopakul, N., Homchan, P., and Thammasiri, K. (2021) Cryopreservation of orchid pollinia using the V cryo-plate. *CryoLetters* 42, 25–32.

Kim, H.H., Yoon, J.W., Ye, P., Cho, E.G., Sohn, J.K., Kim, T.K. and Engelmann, F. (2006) Cryopreservation of potato cultivated varieties and wild species: critical factors in droplet vitrification. *CryoLetters* 27, 223–234.

Klaocheed, S., Rittirat, S., Thammasiri, K. and Prasertsongskun, S. (2018) Alginate-encapsulation, short-term storage and plantlet regeneration from encapsulated protocorm-like bodies (PLBs) of *Cymbidium finlaysonianum* Lindl.: an endangered orchid of Thailand. *Walailak Journal of Science and Technology* 15, 725–737. DOI: 10.48048/wjst.2018.3325

Leunufna, S. and Keller, E.R.J. (2003) Investigating a new cryo-preservation protocol for yams (*Dioscorea* spp.). *Plant Cell Reports* 21, 1159–1166.

Lurswijidjarus, W. and Thammasiri, K. (2004) Cryopreservation of shoot tips of *Dendrobium* Walter Oumae by encapsulation/dehydration. *ScienceAsia* 30, 293–299.

Matsumoto, T. (2017) Cryopreservation of plant genetic resources: conventional and new methods. *Reviews in Agricultural Science* 5, 13–20. DOI: 10.7831/ras.5.13

Matsumoto, T. and Sakai, A. (1995) An approach to enhance dehydration tolerance of alginate-coated dried meristems cooled to -196°C. *CryoLetters* 16, 299–306.

Matsumoto, T., Sakai, A. and Yamada, K. (1994) Cryopreservation of *in vitro*-grown apical meristems of wasabi (*Wasabia japonica*) by vitrification and subsequent high plant regeneration. *Plant Cell Reports* 13, 442–446.

Matsumoto, T., Sakai, A., Takahashi, C. and Yamada, K. (1995) Cryopreservation of *in vitro*-grown meristems of warabi (*Warabia japonica*) by encapsulation-vitrification method. *CryoLetters* 16, 189–196.

Niino, T., Tanaka, D., Tantely, R.R., Fukui, K. and Shirata, K. (2007) Cryopreservation of basal stem buds of in vitro-grown mat rush (*Juncus* spp.) by vitrification. *CryoLetters* 28, 197–206.

Niino, T. and Sakai, A. (1992) Cryopreservation of alginate–coated *in vitro*-grown shoot tips of apple, pear, and mulberry. *Plant Science* 87, 199–206.

Niino, T., Yamamoto, S., Fukui, K., Castillo Martínez, C.R., Valle Arizaga, M., Matsumoto, T. and Engelmann, F. (2013) Dehydration improves cryopreservation of mat rush (*Juncus decipiens* Nakai) basal stem buds on a cryo-plate. *CryoLetters* 34, 549–560.

Niino, T., Watanabe, K., Nohara, N., Rafique, T., Yamamoto, S.I. *et al.* (2014). Cryopreservation of mat rush lateral buds by air dehydration using aluminum cryo-plate. *Plant Biotechnology* 31, 281–287.

Office of Environmental Policy and Planning (OEPP) (1996) Thailand's biodiversity. Available at: https://www.cbd.int/doc/world/th/th-nbsap-01-en.pdf (accessed 10 May 2025).

Panis, B., Piette, B. and Swennen, R. (2005) Droplet vitrification of apical meristems: a cryopreservation protocol applicable to all Musaceae. *Plant Science* 168, 45–55.

Pathirana, R., Mclachlan, A., Hedderley, D., Carra, A., Carimi, F. and Panis, B. (2015) Removal of leafroll viruses from infected grapevine plants by droplet vitrification. *Acta Horticulturae* 1083, 491–498.

Pennycooke, J.C. and Towill, L.E. (2000) Cryopreservation of shoot tips *in vitro* plants of sweet potato [*Ipomoea batatas* (L.) Lam.] by vitrification. *Plant Cell Reports* 19, 733–737.

Pornchuti, W. and Thammasiri, K. (2008) Cryopreservation of protocorms of *Dendrobium cariniferum* Rchb. f. *Acta Horticulturae* 788, 63–68.

Prasongsom, S., Thammasiri, K., Chuenboonngarm, N., Panvisavas, N., Narangajavana, J. and Thitamadee, S. (2020) Conservation of *Dendrobium cruentum* Rchb. f. *Acta Horticulturae* 1298, 187–194.

Rittirat, S., Klaocheed, S. and Thammasiri, K. (2015) Short-term storage of alginate-encapsulated protocorm-like bodies of *Phalaenopsis cornu-cervi* (Breda) Blume & Rchb. f. *Acta Horticulturae* 1078, 75–80.

Rittirat, S., Klaocheed, S., Suppapan, J., Chaithada., P., Kalawong, S. and Thammasiri, K. (2019) Cryopreservation of an endangered pharmaceutically important orchid, *Cymbidium finlaysonianum* Lindl. using vitrification technique. *Acta Horticulturae* 1234, 125–132.

Sakai, A. (1960). Survival of the twigs of woody plants at -196°C. *Nature* 185, 393–394. DOI: 10.1038/185393a0

Sakai, A. and Nishiyama, Y. (1978) Cryopreservation of winter vegetative buds of hardy fruit trees in liquid nitrogen. *HortScience* 13, 225–227.

Sakai, A., Kobayashi, S. and Oiyama, I. (1990) Cryopreservation of nucellar cells of naval orange (*Citrus sinensis* Osb. Var. Brasiliensis Tanaka) by vitrification. *Plant Cell Reports* 9, 30–33.

Sakai, A. (2000) Development of cryopreservation techniques. In: Engelmann, F. and Tagaki, H. (eds) *Proceedings of the International Workshop on Cryopreservation of Tropical Plant Germplasm: Current Research Progress and Application*. Japan International Research Center for Agricultural Sciences, Tsukuba, Japan, pp. 1–7.

Sakai, A. and Englemann, F. (2007) Vitrification, encapsulation-vitrification and droplet-vitrification: a review. *CryoLetters* 28, 151–172.

Sakai, A., Hirai, D. and Niino, T. (2008) Development of PVS-based vitrification and encapsulation-vitrification protocols. In: Reed, B.M. (ed.) *Plant Cryopreservation: A Practical Guide*. Springer, New York, USA, pp. 33–58.

Soonthornkalump, S., Meesawat, U. and Sutthinon, P. (2020) Pollen viability and the preliminary study of pollen preservation method of *Amorphophallus koratensis* Gagnep. (*Araceae*), the potential exotic ornamental plant. *Acta Horticulturae* 1298, 183–186.

Sopalun, K., Thammasiri, K. and Ishikawa, K. (2010) Vitrification-based cryopreservation of *Grammatophyllum speciosum* protocorms. *CryoLetters* 31, 347–357.

Thammasiri, K. (2000) Cryopreservation of seeds of a Thai orchid (*Doritis pulcherrima* Lindl.) by vitrification. *CryoLetters* 21, 237–244.

Thanasuttanithi, D. and Thammasiri, K. (2022) Cryopreservation of *Grammatophyllum speciosum* Blume seeds by D cryo-plate method. *Acta Horticulturae* 1339, 111–116.

Tokuhara, K. and Mii, M. (1993) Micropropagation of *Phalaenopsis* and *Doritaenopsis* by culturing shoot tips of flower stalk buds. *Plant Cell Reports* 13, 7–11.

Towill, L.E. and Bonnart, R. (2003) Cracking in a vitrification solution during cooling or warming does not affect growth of cryopreserved mint shoot tips. *CryoLetters* 24, 341–346.

Towill, L.E. and Forsline, P.L. (1999) Cryopreservation of sour cherry (*Prunus cerasus* L.) using a dormant vegetative bud method. *CryoLetters* 20, 215–222.

Usagami, A., Sakai, A., Nagai, M. and Takahashi, T. (1989) Survival of cultured cells and somatic embryos of *Asparagus officinalis* cryopreserved by vitrification. *Plant Cell Reports* 8, 418–421.

Vacin, E.F. and Went F.W. (1949) Some pH changes in nutrient solutions. *Botanical Gazette* 110, 605–613.

Volk, G.M., Bonnart, R., Krueger, R. and Lee, R. (2012) Cryopreservation of citrus shoot tips using micrografting for recovery. *CryoLetters* 33, 418–426.

Yamamoto, S., Rafique, T., Priyantha, W.S., Fukui, K., Matsumoto, T. and Niino, T. (2011) Development of a cryopreservation procedure using aluminum cryo-plates. *CryoLetters* 32, 256–265.

Zhang, J., Xin, X., Yin, G., Lu, X. and Chen, X. (2014) *In vitro* conservation and cryopreservation in National Genebank of China. *Acta Horticulturae* 1039, 309–331.

5 Tissue Culture of Ornamental Plants in Thailand

Kanchit Thammasiri*

Center for Gardening and Horticulture, Xishuangbanna Tropical Botanical Garden, Chinese Academy of Sciences, Menglun, Mengla, Yunnan, China

Abstract

Plant tissue culture in Thailand is successfully applied to orchids and other ornamentals, which are now a leading agricultural export. The acceptance of tissue culture by farmers and companies was the result of careful design of facilities, and the use of local agricultural resources as substitutes for expensive reagents. Also of great importance was the validation and application of foreign propagation protocols to local orchids. This chapter discusses the key aspects of the layout of a tissue culture laboratory, best practices, and the best culture media for each ornamental crop with the aim of assisting researchers and growers across Thailand, Asia and the world.

Keywords: Ornamental crops, tissue culture, micropropagation, laboratory design and layout, agricultural development, Thailand.

5.1 The Current Status of Tissue Culture in Thailand

The ornamental crops in Thailand are grown for house gardens, landscape design, cut flowers, and potted flowering and ornamental plants, such as trees, shrubs, vines, orchids, palms, ferns, grasses, bamboos, cacti, succulents, annuals, bulbs, and other flowering crops. The production area of Thai ornamental crops was estimated to be about 26,800 acres (Agricultural Extension, 2022). Orchids with the production area of about 5,600 acres (21% of total) rank the highest among the ornamental crops, especially for cut flowers, which are important to the Thai agriculture and economy, followed by jasmine flowers (3,000 acres), lotus (2,070 acres), *Calotropis gigantea* (1,500 acres), marigold (1,130 acres), rose (635 acres), and chrysanthemum (600 acres). The orchid export value started from less than US$1 million more than fifty years ago to about $78.7 million in 2024. It is estimated that 54% of the orchids produced are exported and the remaining 46% are consumed in the domestic market. The export of cut-flower orchids ($62.3 million) still predominates, but the export of potted orchid plants has also been on a rapid increase, reaching about $16.4 million in 2024 (Department of International Trade Promotion, 2025; Department of Customs, 2025) (Fig. 5.1). Apart from orchids, other exported ornamental crops show a combined sale value of about $33 million, including

*Corresponding author: kanchitthammasiri@gmail.com

 Biotechnology of Orchids and Other Ornamentals (Eds P. Bolaños-Villegas and F-C. Chen)

DOI: 10.1079/9781800627895.0005

Sansevieria, Tillandsia, Euphorbia, Aglaonema, Ficus, Amaryllis, Philodendron, Encephalartos, and others (Fig. 5.2).

Plant tissue culture (micropropagation) is one of the most important techniques for growing plant cells under sterile conditions. A suitable medium composition is a specific factor for the growth and morphogenesis of plant species. The medium consists of water, a solid or semi-solid support, macronutrients, micronutrients, vitamins, carbon sources, plant growth regulators or hormones, amino acids, solvents, and perhaps other undefined organic supplements.

The advantages of micropropagation are rapid clonal propagation for plant breeding and research, production of useful metabolites, propagation of pathogen-free plants, propagation for long-term conservation, and easy and cheap exchange of plants for international exchange. The potential is there to produce a significantly greater number of healthy plants in less space, with less labor, and at low cost

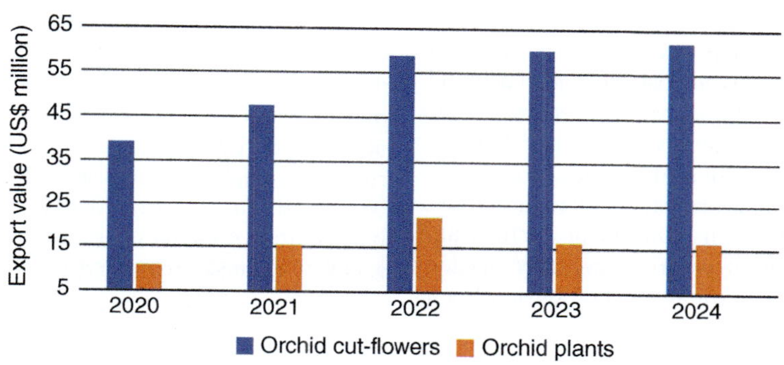

Fig. 5.1. Export value of orchid cut-flowers and plants from Thailand (2020–2024). Reproduced with permission from Department of International Trade Promotion, and Department of Customs, Thailand, 2025.

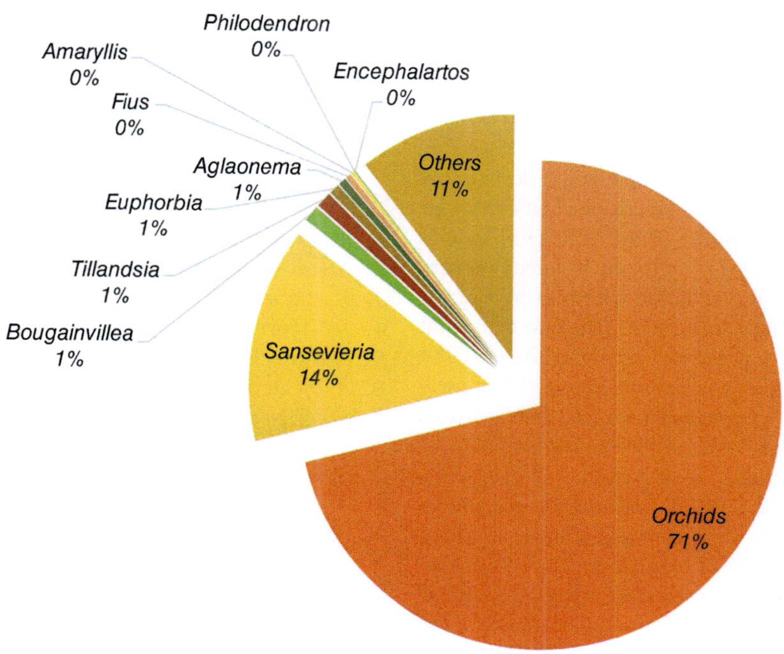

Fig. 5.2. Exports of leading ornamental crop exports from Thailand in 2022, data is shown as percentages. Author's own image.

compared to other means of vegetative propagation (Kyte and Kleyn, 2001).

At present, the biotechnology used for ornamental crops in Thailand is tissue culture and the new development of cultural practices. Molecular techniques are studied only in universities and institutions because the cultivars may change in a short time and returns for any prospective investments will be low. Orchids are a major cash crop that heavily relies on tissue culture due to their popularity, versatility, and environmental sustainability. Other ornamental crops are *Monstera*, *Philodendron*, *Aglaonema*, *Alocasia*, *Tillandsia*, Piperaceae, and Araceae, whose production depends on unstable market demand. Tissue culture techniques have been developed and tested for numerous species and hybrids. Many wild ornamental species are endangered because of deforestation and natural disasters. Therefore, conservation of orchids and other ornamental plants is urgently needed, and these efforts are assisted by setting living collections, seed and pollen storage, *in vitro* conservation, and cryopreservation by government agencies, private companies, and regular people (Thammasiri *et al.*, 2020).

5.2 Ornamental Plant Tissue Culture

Gottlieb Haberlandt in 1902 was the first to attempt the culture of single cells, despite not being successful. He is recognized as the father of plant tissue culture (Gautheret, 1985). Kotte (1922) succeeded in culturing isolated root tips. Success was also achieved with bud cultures by Loo (1945) and Ball (1946). The first true plant tissue cultures were obtained by Gautheret (1934 and 1935) from the cambial tissue of *Acer pseudoplatanus*. The 1940s, 1950s, and 1960s proved to be an exciting time for the development of new techniques.

During the 1990s, continued expansion in the application of *in vitro* technologies to an increasing number of plant species was observed. Tissue culture techniques are being used with all types of plants, including cereals, grasses, etc. Plant tissue culture helps in applied plant biotechnology, as well as stimulating fundamental scientific progress.

5.2.1 Setting up a tissue culture laboratory

A plant tissue culture laboratory has several functional areas, whether it is designed for teaching, research, or business, and no matter what its size or how elaborate it is. Avoid locating it adjacent to laboratories that handle microorganisms or insects, or facilities that are used to store seeds or other plant materials. Contamination from air vents and high people traffic can be a problem. People traffic within a facility scuffs up the wax on floors, as well as disperses dust, which helps to spread contaminants. The tissue culture area should always be kept clean. In designing a plant tissue culture laboratory, three major rooms must be prioritized: the preparation room, the transfer room, and the culture room. In a small laboratory, available space must be divided into two areas, which may be combined into a preparation room and a transfer room, or a transfer room and a culture room.

5.2.2 The preparation room

A media preparation room should have smooth walls and floors that are easy to scrub to maintain a high degree of cleanliness, as well as enough space to prepare the media on the bench with pH meters, analytical balance machines, and a sink nearby. The reagents and stock solutions should be placed alphabetically on shelves or in a refrigerator adjacent to the bench. The equipment used in these rooms is:

1. Analytical balance machines.
2. A hotplate or a microwave.
3. A magnetic stirrer.
4. A pH meter.
5. An autoclave.
6. A hot air oven.
7. A refrigerator.

5.2.3 The transfer room

The transfer room needs to be as clean as possible and have good ventilation. It should be isolated as much as possible from outside doors to prevent contamination, and opening of doors should be minimized. The walls, floor, and ceiling must be painted with light colors, and surfaces should be smooth to ensure frequent cleaning with detergent and standard disinfectants. And the light intensity inside the laminar air-flow cabinet must be sufficient for work. Sterilization of the instruments can be done with glass-bead sterilizers or flaming after dipping in 90% ethyl alcohol. The culture containers should be stacked on mobile carts (trolleys) to facilitate easy movement from the medium storage room to the transfer room, and finally to the culture room. The seats of the transfer operators should be comfortable, as they must work for long periods in the same place. Fire extinguishers and first aid kits should be provided in the transfer room as a safety measure. The personnel should leave their shoes outside the room. Special laboratory shoes and coats should be worn in this area. Ultraviolet (UV) lights are sometimes installed in transfer areas to disinfect the room; these lights should be used only when people and plant material are not in the room. Safety switches can be installed to turn off the UV lights when regular room lights are turned on. The transfer room ideally should contain:

1. A laminar air-flow cabinet with comfortable chairs.
2. A dissecting microscope.
3. Gas outlets.
4. Vacuum lines.
5. Forceps, spatulas, a scalpel, and disposable blades.

5.2.4 The culture room

Controlled temperature, lighting, relative humidity, and shelving need to be considered in planning the culture room. The culture room should have suitable lights for different growth stages and temperature (20–28°C). Air quality should be clean without dust. High humidity in the culture room should be avoided because it increases contamination. Some culture rooms have dehumidifiers and air scrubbers. There is no need for windows except when natural light is used. Overhead light sources can be minimized because the primary source of illumination is from the lights mounted on the shelves. The culture room contains:

1. Shelves with lighting on a timer and controlled temperature.
2. Incubators with controlled temperature and light.
3. Shakers for liquid medium.

5.2.5 Media components and preparation

The media is composed of inorganic compounds, plant growth regulators, vitamins, amino acids and complex organic supplements, carbohydrates, water, gelling agents, and sometimes activated charcoal. Table 5.1 shows components of the commonly used tissue culture medium.

Inorganic compounds are composed of microelements (N, P, K, Ca, Mg, and S), which all plants require in relatively large amounts, and microelements (Fe, Mn, Zn, B, Cu, Mo, and Cl) in relatively small amounts.

Auxins and cytokinins are two main groups of plant growth regulators that are often used in several plants (Wu *et al.*, 2004) and are required for stages of plant tissue culture. Auxins are used for inducing growth of plant roots, stimulating shoot cell elongation and callus formation, or enhancing the effect of cytokinins. Cytokinins are popularly used for shoot proliferation, promoting cell division, and regulating growth and development. The presence of auxins in the medium is generally essential for embryo initiation (Steward *et al.*, 1958). Auxins play two main roles in plant tissue culture. Firstly, they can be used to induce root formation, such as indole acetic acid (IAA) and indole butyric acid (IBA). Secondly, they can initiate callus formation, such as 2,4-dichlorophenoxy acetic acid (2,4-D), naphthalene acetic acid (NAA), and thidiazuron (TDZ) used together with NAA (Chen *et al.*, 2004). TDZ is a substituted phenyl urea with cytokinin-like activity. It is generally believed that TDZ is more active in stimulating shoot formation than 6-benzyladenine (BA) or kinetin, even at extremely low concentrations (Huetteman and Preece, 1993). The disadvantage of TDZ in regeneration is the difficulty in elongation and rooting of the regenerated shoots. This problem could be overcome by transferring regenerated shoots to Murashige and Skoog (MS) (1962) medium supplemented with cytokinin 6-benzyladenine and NAA.

Vitamins have catalytic functions in enzyme systems and are required in trace amounts. Thiamine (vitamin B_1) is essential for all plant

Table 5.1. Components of commonly used tissue culture media (mg l^{-1}) for ornamentals in Thailand. Author's own table.

Components	Murashige and Skoog, 1962	Gamborg *et al.*, 1968	White, 1963	Lloyd and McCown, 1981	Vacin and Went, 1949	Modified Knudson, 1946	Mitra *et al.*, 1976	Nitsch and Nitsch, 1969
Macronutrients								
$Ca_3(PO_4)_2$					200			
NH_4NO_3	1650			400				720
KNO_3	1900	2500	80		525	180	180	950
$CaCl_2 2H_2O$	440	150		96				166
$MgSO_4 7H_2O$	370	250	720	370	250	250	250	185
KH_2PO_4	170			170	250	150	150	68
$(NH_4)_2SO_4$		134			500	100	100	
$NaH_2PO_4.H_2O$		150	16.50					
$Ca(NO_3)_2.4H_2O$			300	556		200	200	
Na_2SO_4			200					
KCl			65					
K_2SO_4				990				
Micronutrients								
KI	0.83	0.75	0.75			80	0.03	
H_3BO_3	6.2	3	1.5	6.2		6.2	0.6	10
$MnSO_4.4H_2O$	22.3		7		0.75	0.075		25
$MnSO_4.H_2O$		10		29.43				
$ZnSO_4.7H_2O$	8.6	2	2.6	8.6			0.05	10
$Na_2Mo0_4.2H_2O$	0.25	0.25		0.25		0.25	0.05	0.25
$CuSO_4.5H_2O$	0.025	0.025		0.25		0.025	0.05	0.025
$CoCl_2.6H_2O$	0.025	0.025				0.025		
$Co(NO_3)_2.6H_2O$							0.05	
Na_2EDTA	37.3	37.3		37.3		74.6	37.3	37.3
$FeSO_4.7H_2O$	27.8	27.8		27.8		25	27.8	27.8
$MnCl_2$						3.9	0.4	
$Fe(C_4H_4O_6)_3.2H_2O$					28			

Vitamins and other supplements								
Inositol	100	100		100				100
Glycine	2	2	3	2				2
Thiamine HCl	0.1	10	0.1	1		0.3	0.3	0.5
Pyridoxine HCl	0.5		0.1	0.5		0.3	0.3	0.5
Nicotinic acid	0.5		0.5	0.5			1.25	5
Ca-pantothenate			1					
Cysteine HCl			1					
Riboflavin						0.3	0.05	
Biotin							0.05	0.05
Folic acid							0.3	0.5

tissue cultures, whereas nicotinic acid (niacin) and pyridoxine (vitamin B_6) may stimulate late growth (Gamborg *et al.*, 1976).

Amino acids (mainly glycine) and complex organic supplements support the growth of tissues. Examples are coconut water, peptone, yeast extract, malt extract, fruit juices, and fruit pulps, and combinations of them are used because of their cheap price and the little labor required to prepare them.

Carbohydrates are an energy and carbon source that are added to all media because most cultures are not completely autotrophic. Carbohydrates are a very important component of *in vitro* cultures because they serve as an energy and carbon source, as well as an osmotic agent. They are added to any nutrient media that is essential for *in vitro* growth and development because photosynthesis is insufficient during *in vitro* culture (Pierik, 1987). Sucrose (20–40 g l^{-1}) is commonly used in plant tissue culture in addition to glucose, fructose, and sorbitol under some conditions.

Water is employed in all media, including the water used during the culture procedure. About 95% of the nutrient medium consists of water. Distilled water is needed for research; otherwise, purified water by reverse osmosis, de-ionization, and/or filtration is appropriate.

Gelling agents are mostly used for stationary supports. Agar, agarose, filter paper, cotton, cheesecloth, vermiculite, etc. are used as gelling agents depending on the purpose and suitability of the culture. Agar is a natural product that is obtained from various types of seaweeds. Agarose is a highly purified linear galactan hydrocolloid isolated from agar or agar-containing seaweed.

Activated charcoal (AC) is composed of carbon aggregates that have numerous pores on the surface. AC was used as a culture component for adsorption of toxic plant metabolites, such as phenolic compounds and plant growth regulators (Thomas, 2008). Moreover, AC may provide a dark environment during *in vitro* culture that influences shoot and root production. For instance, AC may effectively stimulate the formation of somatic embryos on half-strength MS medium (Naing *et al.*, 2011). In general, graphite and AC are odorless, and non-toxic substances; therefore, they can be interchanged to darken the culture medium of plant tissue culture.

5.3 Explant Preparation

The explant is a portion of a plant (tissue, organ, a few cells, or part of a callus mass) excised for culture *in vitro*. Strong and healthy, young, collected in the right season (springtime is appropriate), of large size (larger is better for more nutrients and plant growth regulators), these plants should be selected without viruses, bacteria, fungi, mites, etc. After detaching explants from the source plant, they must be cleaned. There is no rule for cleaning explants; it depends on how clean or dirty the explants are to begin with. The most common cleaning agent used to disinfect explants is bleach in various concentrations (pure household bleach is 5% or 6% sodium hypochlorite). Commonly used concentrations are 5%, 10%, and 20%. Other common cleaning dilutions are 70% ethyl alcohol and 3% hydrogen peroxide. The cleaning should be done in the media preparation room. Transport the cleansed explants to the transfer hood, still in bleach, in a covered container, and apply the final rinses in the laminar flow hood using sterile techniques.

5.3.1 The importance of enforcing an aseptic technique in tissue culture

A sterile environment during the culture of plant tissues must be maintained. Air currents must be avoided over the sterile area because these carry spores of contaminating microorganisms. Laminar air-flow cabinets allow a gentle flow of ultrafiltered sterile air to pass across the working area, thus preventing airborne contamination. Before the start of any sterile procedures, the working area should be thoroughly scrubbed with a tissue soaked with ethyl alcohol (70% v/v). Several techniques are employed for the sterilization of glassware, surgical instruments, liquids, and plant materials. The goal in surface sterilization of plant tissue is to remove all microorganisms with minimum damage to the plant system to be cultured. Methods are dry heat, wet heat, ultrafiltration, and chemical sterilization.

Dry heat is used only for glassware, metal instruments, or other materials that are not charred by high temperatures. Cotton, paper, or plastic cannot be used. Using 160°C (320°F) for 4 hours is sufficient. On the contrary, wet heat is used for liquids, at a steam pressure of 15 lb in^{-2} (103 kPa) from an autoclave with vapor under pressure at 121°C (250°F) for 15 minutes. Do not close the escape valve. At the end of 20 min period, the pressure must be allowed to return slowly to the atmospheric level because rapid decompression will cause the liquids to boil out of the containers. Prolonged autoclave times must be avoided because this results in the decomposition of the chemicals present in the medium.

Ultrafiltration is used for some medium components that are heat-labile and must be sterilized at room temperature, then added to cooling agar medium, and still in the solution state, such as some growth regulators. We suggest the use of either sintered glass filters, or nucleopore filters. Chemical sterilization on the other hand is also employed during tissue culture, such as in the case of 70% ethyl alcohol for surface-sterilization, or in the case of 0.5% NaO-Cl (sodium hypochlorite), a sterilizing solution for general use.

Cleanliness is of primary importance in the transfer room. The following practices should always be observed to maintain safe and clean conditions in the laboratory:

1. No eating, drinking, smoking, or storing food in the laboratory.
2. Wear a protective lab coat or apron when working with cultures. Long, full sleeves should be avoided because they collect dust.
3. Wear sterilized disposable medical gloves or plastic garden gloves.
4. Tie back and cover long hair. Loose hair is a potential contaminant for cultures and risks catching fire from any open flames.
5. Sit straight and keep your head out of the laminar air-flow cabinet.
6. Set only the necessary equipment in the laminar air-flow cabinet.
7. Sterilize in an autoclave or pressure cooker any jars and petri dishes containing contaminated cultures before emptying them out for disposal.
8. Disinfect work surfaces before and after working in the laminar air-flow cabinet.
9. Keep transfer room doors closed to avoid unnecessary air currents.

5.3.2 Sterile technique for explants

Procedure:

1. Before starting to work, turn on the UV lights for 15 minutes.
2. When preparing the explants or any work, turn on the laminar air-flow cabinet for 10 minutes before using.
3. Wipe both hands and all things, including plant material, with a cloth soaked with 70% ethyl alcohol, then put them inside the cabinet.
4. Wipe or spray inside the cabinet with 70% ethyl alcohol. Do not spray the HEPA filter.
5. Put knives, forceps, and other equipment in a glass bead sterilizer or a container if you use an alcohol lamp or gas for sterilization.
6. Flame the glass Petri dish and open the lid.
7. Place the plant material in the Petri dish.
8. Soak the knife and forceps in 90% ethyl alcohol and flame (if using alcohol lamp or gas).
9. Cut the plant material you will culture.
10. Flame the prepared medium container lid and open it.
11. Place the explant on the medium.
12. Flame the lid and the edge of the medium container, then close the lid.
13. After finishing, take the containers out and then wipe the inside cabinet and close the cabinet.

5.3.3 Sterile technique for transfer

Procedure:

1–5. They are the same as in the sterile technique for explants.
6. Flame both a medium container with explant and a new medium container and open both lids.

7. Place the explant on the new medium.
8. Flame the lid and the edge of the new medium container, then close the lid.
9. After finishing, take the containers out and then wipe the inside cabinet and close the cabinet.

5.4 How to Prevent and Handle Contamination

One of the most important factors in preventing contamination of the cultures is the removal of airborne contaminants in the sterile area. Always keep all windows and doors closed to avoid airborne dispersal of contaminants. A minimum number of persons should be present. There are four sources of contamination: (i) the plant (internal as well as external);(ii) the medium, which should be sufficiently sterilized; (iii) the air; and (iv) the person who works. The most important factor is the explant, which must be well sterilized before culture *in vitro*.

Contamination in tissue culture can be expensive, especially if it exceeds 1%. If contamination appears, sterilize the containers before discarding their contents.

5.5 Developmental Pathways in Cultured Cells and Tissues

Roots, shoots, and flowers are the organs that may arise *de novo* from plant tissue culture (organogenesis). Embryos are not classified as organs because these structures have an independent cell fate. Skoog and Miller (1957) observed that a relatively high auxin:cytokinin ratio induced root formation in tobacco calli, whereas a low ratio favored shoot production. Somatic embryogenesis is the production of embryos not from the development of the fertilized egg. Anthers and pollen are cultured to produce haploid homozygous plants by the induction of embryogenesis or organogenesis.

The *in vitro* growth and development of a plant is determined by several complex factors which are: the genetic make-up of the plant, nutrients, physical growth factors (light, temperature, pH, O_2, and CO_2) and some organic substrates (plant growth regulators, vitamins, amino acids, coconut water, banana, etc.).

There are four stages of plant tissue culture growth:

- **Stage I Establishment:** Preparing the explants and starting the culture.
- **Stage II Multiplication:** Proliferation of shoots.
- **Stage III Rooting:** Root initiation.
- **Stage IV Acclimatization or transferring to the greenhouse:** Lowering relative humidity and increasing light intensity, and temperature.

It is not necessary that each stage needs a different culture medium. A single medium may be satisfactory for the first three stages. Stage III can be skipped since rooting may stop multiplication. Instead, rooting will occur at Stage IV.

5.6 Acclimatization of Plantlets

Acclimatization is defined as the climatic or environmental adaptation of an organism, especially a plant, that has been moved to a new environment. Acclimatization is required because there is a significant difference between the tissue culture or micropropagation environment and the greenhouse or field environment. In the laboratory, the cultures have low light intensity, low temperature, 90–100% relative humidity, and cuticle poorly developed.

Good techniques for acclimatizing tissue-cultured plantlets should be applied. Acclimatization can be started while plantlets are still *in vitro* by providing higher light intensity and reducing relative humidity by loosening the lid (closure) for 1–2 weeks before moving the plantlets to the greenhouse.

An abrupt change to the lower humidity and greater light in the greenhouse can be fatal to plantlets in a very short time (Kyte and Kleyn, 2001). In the greenhouse, plantlets should be covered with plastic, watered very often, and gradually increased light intensity for a few weeks before growing normally in the greenhouse.

5.7 How to Succeed at Commercial Production

Successful micropropagation on a commercial scale for producing many quality plants largely depends on high multiplication from suitable protocols and successful acclimatization. It is not enough to produce high-quality, healthy plants; one must also produce them in a timely fashion, giving customers what they want, when they want it, for a price they can afford.

It is desirable to predict the financial success of a commercial laboratory, record it in progress, and periodically evaluate it in retrospect. Good management involves constant planning, study, observation, review, and adjustment. Good marketing practices are essential for success (Kyte and Kleyn, 2001).

Production technology, breeding, tissue culture, and molecular studies in Thailand started from tissue culture research in orchids in some universities in 1967. Later in 1972, the orchid tissue culture business expanded rapidly to over 30 million plantlets and employed over 300 workers in 15 labs with about $2.4 million. In addition, tissue culture protocols for rapid propagation were developed for other ornamental crops, such as *Monstera*, *Philodendron*, *Aglaonema*, *Alocasia*, *Tillandsia*, Piperaceae, Araceae, etc., of which production depends on unstable and erratic demand. Suitable protocols and low-cost tissue culture were developed for commercialization, as well as a conventional breeding program for outstanding cultivars for international markets by the government organizations, institutions, growers, and private companies. Major growing factors in natural growing habitats are studied for developing new cultural practices, such as low-cost greenhouse, watering, fertilizer application, pest and disease control, postharvest handling, packaging, marketing, and logistics for new cultivars and tissue-cultured plantlets.

Tissue culture is an indispensable tool for the commercial production of elite plant selections because of low cost, uniformity, quick propagation, and high yield in a short period (Arditti and Ernst, 1993). It has been successfully carried out for many Thai orchid species and many ornamental plants (Table 5.2 and Fig. 5.3). Most cut-flower orchids, *Dendrobium*, *Oncidium*, *Mokara*, *Aranda*, *Ascocenda*, and those from the *Cattleya* alliance are propagated successfully through tissue culture. Within 1–2 years, one young pseudobulb multiplies to over ten thousand plants from the laboratory and tissue cultured plantlets are ready to be grown in the greenhouse.

5.8 How to Choose the Best Culture Medium for Your Plants

The choice of medium depends mainly on plant species, the tissue, or organ to be cultured, and the purpose. It is best to start with published protocols. A starting point for tissue culture from a dicot tissue explant would be the preparation of the MS medium at different concentrations. MS medium is the most used tissue culture medium and many iterations have been developed. The medium is derived from White's medium and originally developed for the cultivation of *Nicotiana tabacum* calli. Compared to the White medium, the concentration of all components is higher. Due to the high concentration of minerals, the MS medium is a very rich medium that can be too salty for certain plant species. To avoid this problem, MS medium is often used with the microelements in full concentration, but with the macroelements in half or three-quarters of the concentration as originally described. Monocot cultures respond better to the Schenk and Hildebrandt (SH) (1972) medium.

5.9 Conclusions

Thai ornamental crop cultivation for the international markets has a bright future. The export values are high and quite stable. Orchids will continue to dominate other ornamental crops due to better technological know-how in cultivation, suitable climatic conditions, experienced and skillful growers and exporters, as well as their nationwide popularity (Thammasiri, 1997). The orchid cultivation in Thailand is a good example of biotechnology employed for development for an ornamental crop that does not fall in the category of staple food, which rose to become the top crop in the country. Micropropagation was slow to gain acceptance but eventually helped farmers increase income and boost the agrarian economy, bringing prosperity to society as a whole.

Table 5.2. Suggested medium culture and tissue source for the micropropagation of ornamentals in Thailand. Author's own table.

Family	Species	Explant	Regenerant	Medium	References
Acanthaceae	*Staurogyne repens*	Nodal segments	Shoots	Murashige and Skoog (MS) medium + 0.15 mg l^{-1} NAA + 1.0 mg l^{-1} BA	Rittirat *et al.* (2022a)
		Shoots	Roots	MS medium + 3.0 mg l^{-1} IBA	
Apocynaceae	*Adenium obesum* (Forssk.) Roem. and Schult.	Shoot tips	Shoots	MS medium + 22.2 μM BA	Kanchanapoom *et al.* (2010)
Araceae	*Aglaonema tenuipes* Engl.	Seeds	Callus	MS medium + 3.0 mg l^{-1} NAA	Tilarux *et al.* (2022)
		Shoots	Shoots	MS medium + 2.0 mg l^{-1} BA	
		Shoots	Roots	MS medium + 2.0 mg l^{-1} NAA	
	Aglaonema simplex	Apical buds	Shoots	MS medium + 2.0 mg l^{-1} BA	Laohavisuti and Mitrnoi (2005)
		Shoots	Roots	½ MS medium + 1.0 mg l^{-1} IBA	
	Anthurium andraeanum 'HC 028'	Shoot tips	Shoots	MS medium + 4.0 mg l^{-1} BA	Rittirat *et al.* (2022b)
		Shoots	Roots	MS medium + 2.0 mg l^{-1} IBA	
	Anubias barteri var. *nana*	Shoot tips	Plantlets	MS medium + 7.0 mg l^{-1} TDZ	Rittirat *et al.* (2022c)
		Shoot tips	Shoots	MS medium + 7.0 mg l^{-1} kinetin	Rittirat *et al.* (2023)
		Shoots	Roots	½ MS medium	
	Anubias heterophylla	Shoot tips	Shoots	MS medium + 3.0 mg l^{-1} BA	Rittirat *et al.* (2020)
		Shoot tips	Plantlets	MS medium + 0.1% (w/v) AC +1.0 mg l^{-1} NAA + 1.0 mg l^{-1} BA	Rittirat *et al.* (2021a)
	Cryptocoryne wendtii	Shoot tips	Shoots	MS medium + 1 mg l^{-1} NAA	Choojun *et al.* (2017)
		Shoot tips	Multiple shoots	MS medium + 3.0 mg l^{-1} BA	Klaocheed *et al.* (2020)
		Shoots	Roots	MS medium + 1.0 mg l^{-1} NAA	
	Cryptocoryne walkerlii	Shoot tips	Shoots	MS medium + 3.0 mg l^{-1} kinetin	Rittirat *et al.* (2018a)
	Philodendron billietiae	Shoot tips	Shoots	MS medium + 3.0 mg l^{-1} BA	Maikaeo *et al.* (2024)
		Shoots	Roots	½ MS medium + 2.0 mg l^{-1} NAA	
Campanulaceae	*Lobelia cardinalis* L.	Nodal segments	Multiple shoots	MS medium + 1.0 mg l^{-1} BA	Rittirat *et al.* (2021b)
		Micro shoots	Roots	MS medium + 0.5 mg l^{-1} NAA	Rittirat *et al.* (2022d)
	Lobelia cardinalis L.	Micro shoots	Roots	MS medium + 1.0 mg l^{-1} IBA	Rittirat *et al.* (2022e)
	Lobelia cardinalis L.	Nodal segments	Multiple shoots	MS medium + 1.0 mg l^{-1} BA + 2.0 mg l^{-1} NAA	Klaocheed *et al.* (2024b)
		Shoots	Roots	MS medium + 1.0 mg l^{-1} IBA	
Dioscoreaceae	*Tacca chantrieri* Andre.	Shoots	Multiple shoots	MS medium + 3.0 mg l^{-1} BA	Rittirat *et al.* (2018b)

Orchidaceae	*Aerides odorata* Lour.	Protocorms	PLBs	MS + 2.5 mg l^{-1} BA	Rittirat *et al.* (2019)
		Shoots	Plantlets	MS medium + 3.0 mg l^{-1} IBA	
		Protocorms	Plantlets	MS medium + 15% (v/v) Coconut water (CW)+ 0.2 % (w/v) AC	Rittirat *et al.* (2022f)
	Hybrid *Cattleya*	Shoots (1.0-1.5 cm)	Roots	MS medium + 15% (v/v) CW + 3.0 mg l^{-1} NAA	Rittirat *et al.* (2021c)
	Cymbidium finlaysonianum Lindl.	Protocorm	PLBs	Vacin and Went (VW) liquid medium + 8.84 µM BA	Rittirat *et al.* (2017)
	Cymbidium finlaysonianum Lindl.	PLBs	Plantlets	VW medium + 0.2% (w/v) AC	Rittirat *et al.* (2017)
	Cymbidium finlaysonianum Lindl.	Seeds	Protocorms	VW medium	Rittirat *et al.* (2018c)
		Protocorms	PLBs	VW liquid medium + 8.84 µM BA	
		PLBs	Plantlets	VW medium + 0.2 % (w/v) AC	
	Dendrobium cruentum Rchb. f.	Shoot tips	PLBs	½ MS liquid medium + 1% (w/v) sucrose + 1.0 mg l^{-1} NAA	Sangdum *et al.* (2017)
Orchidaceae	*Dendrobium crumenatum* Sw.	Callus	PLBs	MS medium + 0.5 mg l^{-1} TDZ	Rittirat *et al.* (2018d)
		PLBs	Plantlets	MS medium + 1.5 mg l^{-1} IBA	
	Dendrobium crumenatum Sw.	PLBs	Shoots	MS + 15 % (v/v) CW	Klaocheed *et al.* (2021)
		Shoots	Plantlets	MS + 15 % (v/v) CW + 0.2 % (w/v) AC	
	Dendrobium palpebrae Lindl.	Protocorms	Plantlets	VW medium + 7% (w/v) sucrose	Kalawong *et al.* (2020)
	Grammatophyllum speciosum	Shoot tips	PLBs	½ MS liquid medium + 2% (w/v) sucrose	Sopalun *et al.* (2010)
	Grammatophyllum speciosum	Protocorms (2.0 mm)	Shoots	MS liquid medium + 2 mg l^{-1} BA	Samala *et al.* (2020)
	Phalaenopsis cornu-cervi (Breda) Blume & Rchb. f.	Wounded protocorm	PLBs	½ MS medium + 0.1 mg l^{-1} NAA + 0.1 mg L^{-1} TDZ	Rittirat *et al.* (2012a)
	Phalaenopsis cornu-cervi (Breda) Blume & Rchb. f.	PLBs	Plant conversion	MS medium + 15% CW + 0.2% AC	Rittirat *et al.* (2012b)
	Phalaenopsis cornu-cervi (Breda) Blume & Rchb. f.	Leaf explants	PLBs	½ MS medium + 9 µM TDZ	Rittirat *et al.* (2014)
	Rhynchostylis gigantea (Lindl.) Ridl.	Shoot tips	Adventitious shoots	VW liquid medium + 10 g l^{-1} Sucrose	Prasongsom *et al.* (2016)

Continued

Table 5.2. Continued.

Family	Species	Explant	Regenerant	Medium	References
Orchidaceae	*Rhynchostylis gigantean* var. Sagarik	embryogenic callus	somatic embryo (SE)	New Dogashima Medium (NDM) medium + 2% sucrose + 15% CW + 0.2% (w/v) AC + under light condition	Rittirat *et al.* (2011)
	Spathoglottis eburnea Gagnep	Shoot tips	Shoots	½ MS medium + 1 mg l^{-1} NAA + 2 mg l^{-1} BA	Pornchuti *et al.* (2017)
	Spathoglottis plicata Blume	Protocorms	PLBs	MS medium + 15% (v/v) CW + 3.0 mg l^{-1} BA	Rittirat *et al.* (2024)
		PLBs	Plantlets	MS medium + 0.2% (w/v) AC	
	Vanda coerulea	Shoot tips	Adventitious shoots	VW medium + 10 g l^{-1} sucrose	Jitsopakul *et al.* (2013)
Oxalidaceae	*Oxalis triangularis* A.st.-Hil	Petiole segments	Plantlets	MS medium + 3.0 mg l^{-1} BA + 0.5 mg l^{-1} 2,4-D.	Rittirat *et al.* (2022g)
	Oxalis triangularis A.st.-Hil	Petiole segments	Shoots	MS medium + 0.5 mg l^{-1} NAA + 1.0 mg l^{-1} BA	Klaocheed *et al.* (2024)
		Shoots	Roots	½ MS medium + 1.0 mg l^{-1} IBA	
Zingiberaceae	*Globba schomburgkii* Hook. f. via Bulbil	Microshoots (1 cm)	Shoots	MS medium + 2 mg l^{-1} TDZ	Saensouk *et al.* (2018)
	Kaempferia koratensis Picheans.	Microshoots (1 cm)	Shoots	Liquid MS medium + 2 mg l^{-1} TDZ + 1 mg l^{-1} kinetin + 0.2 mg l^{-1} NAA	Saliwan *et al.* (2022)
	Kaempferia parviflora Wall. ex Baker	Terminal buds	Shoots	MS medium + 35.52 μM BA	Prathanturarug *et al.* (2007)
	Kaempferia marginata Carey ex Roscoe	Microshoots	Shoots	MS medium + 2 mg l^{-1} BA and 1.0 mg l^{-1} TDZ	Saensouk *et al.* (2016)
	Curcuma lithophila Škorničk. & Soonthornk.	Microshoots (1.5 cm)	Shoots	MS medium + 10 μM BA	Nakdang *et al.* (2023)
	Curcuma longa Linn.	Rhizomes	Callus	MS medium + 0.5 mg l^{-1} 2,4-D	Kaewthip *et al.* (2021)
	Curcuma comosa Roxb.	Bud	Shoots	MS medium + 3.0 mg l^{-1} BA	Siringam and Pongprayoon (2022)
	Curcuma singularis Gagnep.	Rhizome buds	Shoots and roots	MS medium + 2.0 mg l^{-1} BA	Jitsopakul *et al.* (2017)
	Kaempferia galanga Linn.				
	Zingiber officinale Rosc.	Rhizome buds	Shoots and roots	MS medium + 1.0 mg l^{-1} BA	
	Kaempferia angustifolia	Bud	Shoots	MS medium + 1.0 mg l^{-1} kinetin + 2.0 mg l^{-1} TDZ + 0.2 mg l^{-1} NAA	Saliwan *et al.* (2022)

Fig. 5.3. Tissue culture protocol of *Grammatophyllum speciosum*. A) A shoot tip was excised under a stereo microscope. B) Protocorm-like bodies (PLBs) were induced in half-strength MS liquid medium (2 months after culture). C) PLBs were subcultured for multiplication (1 month after culture). D) PLBs were cultured on half-strength MS solid medium supplemented with 0.05% (w/v) of activated charcoal, 2.0 mg l^{-1} NAA, and 1.0 mg l^{-1} BA. E and F) Shoots and roots were induced (3 months after culture). Sopalun *et al*., 2010. Reproduced with permission from Springer Science Business Media B.V.

References

Agricultural Extension Office of Thailand (2022) Data on ornamental plant production area in Thailand. Available at: https://www.doae.go.th/en/academic-media-information/ (accessed 4 May 2025).

Arditti, J. and Ernst, R. (1993) Micropropagation of Orchids. John Wiley & Sons, New York, USA.

Ball, E (1946) Development in sterile culture of stems tips and subjacent regions of *Tropaeolum majus* L. and of *Lupinus albus* L. *American Journal of Botany* 33, 301–318.

Chen, T.Y., Chen, J.T. and Chang, W.C. (2004) Plant regeneration through direct shoot bud formation from leaf cultures of *Paphiopedilum* orchids. *Plant Cell Tissue and Organ Culture* 76, 11–15.

Choojun, W., Chesu, W., Rittirat, S., Klaocheed, S. and Thammasiri, K. (2017) Micropropagation of an Ornamental Aquatic Plant, *Cryptocoryne wendtii* From Shoot Tip Culture. Proceedings of the 1st national and international conference on education for sustainable locality development 2017. Bansomdejchaopraya Rajabhat University, Bangkok, Thailand, pp. 558–564.

Department of Customs of Thailand (2025) Data on import-export of ornamental crops. Available at: https://rb.gy/83hz0k (accessed 4 May 2025).

Department of International Trade Promotion of Thailand. 2025. Data on import-export of orchids. Available at: https://www.ditp.go.th/post/1166 (accessed 4 May 2025).

Gamborg, O.L., Murashige, T., Thorpe, T.A. and Vasil, I.K. (1976) Plant tissue culture media. *In Vitro* 12, 473–478.

Gautheret, R.J. (1934) Culture du tissus cambial. *Comptes Rendus Hebdomadaires des Séances de l'Académie des Sciences* 198, 2195–2196.

Gautheret, R.J. (1935) Recherches sur la culture des tissus végétaux. PhD Thesis. Science Faculty of Paris, France.

Gautheret, R.J. (1985) History of plant tissue culture: a personal account. In: Vasil, I.K. (ed.) *Cell Culture and Somatic Cell Genetics of Plants Volume 2*. Academic Press, New York, USA, pp. 1–59.

Haberlandt, G. (1902) Kulturversuche mit isolierten Pflanzenzellen (Culture experiments with isolated plant cells). *Sitzungsberichte der Kaiserlichen Akademie der Wissenschaften. Mathematisch-naturwissenschaftliche Classe, Abteilung I* (Proceedings of the Imperial Academy of Sciences. Mathematical and Natural Sciences Class. Department I /Austria): 69–92.

Huetteman, C.A. and Preece, J.E. (1993) Thidiazuron: a potent cytokinin for woody plant tissue culture. *Plant Cell Tissue Organ Culture* 39, 105–119.

Jitsopakul, N., Thammasiri, K. and Ishikawa, K. (2013) Efficient adventitious shoot regeneration from shoot tip culture of *Vanda coerulea*, a Thai orchid. *ScienceAsia* 39, 449–455.

Jitsopakul, N., Sangyojarn, P., Homchan, P. and Thammasiri, K. (2017) Micropropagation for conservation of *Zingiberaceae* in Surin province, Thailand. *Acta Horticulturae* 1167, 75–80.

Kaewthip, W., S. Dheeranupattana, S., Junta, P. and Shank, L. (2021) Sterile tissue preparation and callus induction of *Curcuma longa* Linn. *Chiang Mai University Journal of Natural Sciences* 20: e2021062.

Kalawong, S., Yimyong, P., Rittirat, S., Suwanno, S. and Siangsuepchart, A. (2020) Effects of culture media, sucrose and chitosan concentrations on *Dendrobium palpebrae* Lindl. micropropagation. *Khon Kaen Agriculture Journal* 48, 395–404.

Kanchanapoom, K., Sunheem, S. and Kanchanapoom, K. (2010) *In vitro* propagation of *Adenium obesum* (Forssk.) Roem. and Schult. *Notulae Botanicae Horti Agrobotanici Cluj-Napoca* 38, 209–213. DOI: 10.15835/nbha3834604

Klaocheed, S., Jehsu, W., Choojun, W., Thammasiri, K., Prasertsongskun, S. and Rittirat, S. (2020) Induction of direct shoot organogenesis from shoot tip explants of an ornamental aquatic plant, *Cryptocoryne wendtii*. *Walailak Journal of Science and Technology* 17, 293–302. DOI: 10.48048/wjst.2020.3706

Klaocheed, S., Rittirat, S., Kalawong, S. and Thammasiri, K. (2024a) Rapid clonal multiplication through *in vitro* axillary shoot proliferation of *Lobelia cardinalis* L., an ornamental aquatic plant. *Trends in Sciences* 21, 7690–7690. DOI: 10.48048/tis.2024.7690

Klaocheed, S., Rittirat, S., Kalawong, S. and Thammasiri, K. (2024b) Maximization of micropropagule production in an attractive ornamental plant, *Oxalis triangularis* A.st.-Hil. *Trends in Sciences* 21, 7917. DOI: 10.48048/tis.2024.7917

Klaocheed, S., Rittirat, S. and Thammasiri, K. (2021) Plantlet regeneration and multiple shoot induction from Protocorm-Like Bodies (PLBs) of medicinal orchid species, *Dendrobium crumenatum* Sw. *Walailak Journal of Science and Technology* 18, 9168. DOI: 10.48048/wjst.2021.9168

Kotte, W. (1922) Kulturversuche mit isolierten Wurzelspitzen (culture experiments with isolated root tips). *Beiträge zur Allgemeinen Botanik* (Contributions to General Botany/Germany) 2, 413–434.

Kyte, L. and Kleyn, J. (2001) *Plants from Test Tubes*. Timber Press, Oregon, U.S.A.

Laohavisuti, N. and Mitrnoi, M. (2005) Micropropagation of *Aglaonema simplex*. Proceedings of the 43rd Kasetsart University Annual Conference: Fisheries, Natural Resources and Environmental Economics. Bangkok, Thailand, pp. 267–274.

Loo, S.W. (1945) Cultivation of excised stem tips of *Asparagus in vitro*. *American Journal of Botany* 32, 13–17. https://www.jstor.org/stable/2446523 (accessed 4 May 2025).

Maikaeo, L., Puripunyavanich, V., Limtiyayotin, M., Orpong, P. and Kongpeng, C. (2024) Micropropagation and gamma irradiation mutagenesis in *Philodendron billietiae*. *Thai Journal of Agricultural Science* 57, 11–19. https://li01.tci-thaijo.org/index.php/TJAS/article/view/258326

Murashige, T. and Skoog, F. (1962) A revised medium for rapid growth and bioassays with tobacco tissue culture. *Physiologia Plantarum* 15, 473–497. DOI: 10.1111/j.1399-3054.1962.tb08052.x

Naing, A.H., Chung, J.D. and Lim, K.B. (2011) Plant regeneration through indirect somatic embryogenesis in *Coelogyne cristata* orchid. *American Journal of Plant Science* 2, 262–267. DOI: 10.4236/ajps.2011.22028

Nakdang, S., Chuengpanya, R., Sornyotha, P., Muangkroot, A., Jenjittikul, T. and Chuenboonngarm, N. (2023) *In vitro* propagation of *Curcuma lithophila* Škorničk. & Soonthornk., an endemic species of Thailand with high potential to be used as ornamental plant. *Journal of Applied Science and Emerging Technology* 22: e248222. https://ph01.tci-thaijo.org/index.php/JASCI/article/view/248222

Pierik, R.L.M. (1987) In Vitro *Culture of Higher Plants*. Martinus Nijhoff Publishers, Dordrecht, The Netherlands.

Pornchuti, W., Thammasiri, K., Chuenboonngarm, N. and Panvisavas, N. (2017) Micropropagation of *Spathoglottis eburnea* Gagnep, a Thai orchid species, through shoot tips. *Acta Horticulturae* 1167, 87–94.

Prasongsom, S., Thammasiri, K. and Chuenboonngarm, N. (2016) Efficient adventitious shoot regeneration from shoot tip culture of *Rhynchostylis gigantea* (Lindl.) Ridl.(amethyst-purple), a rare Thai orchid species. *Walailak Journal of Science and Technology* 13, 757–767. https://wjst.wu.ac.th/index.php/wjst/article/view/1921 (accessed 4 May 2025).

Prathanturarug, S., Apichartbutra, T., Chuakul, W. and Saralamp, P. (2007) Mass propagation of *Kaempferia parviflora* Wall. ex Baker by *in vitro* regeneration. *The Journal of Horticultural Science and Biotechnology* 82, 179–183. DOI: 10.1080/14620316.2007.11512217

Rittirat, S., Te-chato, S., Kerdsuwan, N. and Kongruk, S. (2011) Micropropagation of Chang Daeng (*Rhynchostylis gigantea* var. Sagarik) by embryogenic callus. *Songklanakarin Journal of Science & Technology* 33, 659–663. https://www.thaiscience.info/journals/Article/SONG/10891290.pdf (accessed 4 May 2025).

Rittirat, S., Kongruk, S. and Te-chato, S. (2012) Induction of protocorm-like bodies (PLBs) and plantlet regeneration from wounded protocorms of *Phalaenopsis cornu-cervi* (Breda) Blume & Rchb. f. *Journal of Agricultural Technology* 8, 2397–2407. http://www.ijat-aatsea.com/pdf/v8_n7_12_December/21_IJAT_2012_8(7)_Suphat%20Rittirat_Sompong%20Te-chato_plant%20science.pdf (accessed 4 May 2025).

Rittirat, S., Klaocheed, S. and Thammasiri, K. (2014) Enhanced efficiency for propagation of *Phalaenopsis cornu-cervi* (Breda) Blume & Rchb. F. using trimmed leaf technique. *International Journal of Biological, Food, Veterinary and Agricultural Engineering* 8, 328–331.

Rittirat, S., Klaocheed, S., Prasertsongskun, S. and Thammasiri, K. (2017) Asymbiotic seed germination and protocorm-like body proliferation of *Cymbidium finlaysonianum* Lindl. *Acta Horticulturae* 1167, 101–106.

Rittirat, S., Thammasiri, K. and Klaocheed, S. (2018a). Plantlet Production from Shoot Tip Explants of Aquatic Plant, *Cryptocoryne walkerlii* Cultured *In Vitro*. Proceedings of the International Conference on Innovation, Smart Culture and Well-beings (ICISW2018). Suan Sunandha Rajabhat University, Bangkok, Thailand, pp. 370–378. http://conference.ssru.ac.th/download/processding2.pdf (accessed 4 May 2025).

Rittirat, S., Thammasiri, K. and Klaocheed. S. (2018b) *In Vitro* Propagation of an Endangered Medicinal Plant, Bat Flower; *Tacca Chantrieri* Andre. for Conservation in Thailand. Proceedings of the International Conference on Innovation, Smart Culture and Well-beings (Icisw2018). Suan Sunandha Rajabhat University, Bangkok, Thailand, pp. 387–393. http://conference.ssru.ac.th/download/processding2.pdf (accessed 4 May 2025).

Rittirat, S., Klaocheed, S., Thammasiri, K. and Prasertsongskun, S. (2018c) *In vitro* propagation and forest reestablishment of *Cymbidium finlaysonianum* Lindl., an endangered medicinal orchid. *Walailak Journal of Science and Technology* 15, 711–724. http://wjst.wu.ac.th/index.php/wjst/article/view/3326/932 (accessed 4 May 2025).

Rittirat, S., Thammasiri, K., Kalawong, S. and Klaocheed, S. (2018d) *Ex-situ* Conservation of Endangered Medicinal Orchid Species, *Dendrobium crumenatum* Sw. by Micropropagation. The 13th International Conference in Collaboration Between Chiang Rai Rajabhat University, Thailand & University of Miyazaki, Japan, Asian Community Knowledge Networks for the Economy, Society, Culture, and Environmental Stability. Miyazaki, Japan, pp. 46–52.

Rittirat, S., Thammasiri, K. and Klaocheed, S. (2019) Optimization of Plantlet Regeneration from *Aerides odorata* Lour. Proceedings of the 13th Asia Pacific Orchid Conference 2019 (APOC 13), Cultural Exchange Through Orchids. Sarawak, Malaysia, pp. 166–170.

Rittirat, S., Klaocheed, S. Thammasiri, K. and Walam, A. (2020) An efficient *in vitro* plantlet regeneration of an ornamental aquatic plant *Anubias heterophylla* through shoot tip culture. *Acta Horticulturae* 1298, 265–272.

Rittirat, S., Thammasiri, K. and Klaocheed, S. (2021a) *In vitro* rapid multiplication of a highly valuable ornamental aquatic plant *Anubias heterophylla*. *Trends in Sciences* 18, 3. DOI: 10.48048/tis.2021.3

Rittirat, S., Klaocheed, S. and Thammasiri, K. (2021b) *In vitro* shoot multiplication of *Lobelia cardinalis* L., an aquatic ornamental plant. *Acta Horticulturae* 1324, 55–62.

Rittirat, S., Klaocheed, S., and Thammasiri, K. (2021c) *In Vitro* Plantlet Regeneration of Yellow Flowered Hybrid *Cattleya* (Orchidaceae). Proceedings of the 10th International Molecular Biology and Biotechnology Congress (MolBiotech 2021). Çukurova University, Turkey, pp. 49–55.

Rittirat, S., Klaocheed, S. and Thammasiri, K. (2022a) A reliable protocol for micropropagation of an ornamental aquatic plant, *Staurogyne repens*. *Acta Horticulturae* 1339, 227–236.

Rittirat, S., Klaocheed, S. and Thammasiri, K. (2022b) *In vitro* shoot organogenesis in *Anthurium andraeanum* 'HC 028' – a highly valued cut-flower and potted ornamental plant. *Acta Horticulturae* 1339, 291–299.

Rittirat, S., Klaocheed, S. and Thammasiri, K. (2022c) Thidiazuron induces high-frequency plant regeneration from shoot tip explants of an ornamental aquatic plant, *Anubias barteri* var. *nana*. *Acta Horticulturae* 1339, 273–280.

Rittirat, S., Klaocheed, S. and Thammasiri, K. (2022d) Organogenesis and efficient *in vitro* plantlet regeneration from nodal segments of an ornamental aquatic plant, *Lobelia cardinalis* L., using BAP. *Acta Horticulturae* 1339, 237–243.

Rittirat, S., Thammasiri, K. and Klaocheed, S. (2022e) *In Vitro* Shoot Micropropagation and Plant Establishment of an Ornamental Aquatic Pant (*Lobelia cardinalis* L.). Proceedings of the 11th International Molecular Biology and Biotechnology Congress (MolBiotech 2022). Çukurova University, Turkey, pp. 62–68.

Rittirat, S., Klaocheed, S. and Thammasiri, K. (2022f) Asymbiotic *In Vitro* Seed Germination, Plantlet Development and Forest Reestablishment of *Aerides odorata* Lour., an Endangered Orchid from Thailand. Proceedings of the XII International Scientific Conference, Conservation and Cultivation of Orchids. Moscow State University, Moscow, Russia, pp. 224–232.

Rittirat, S., Klaocheed, S. and Thammasiri, K. (2022g) Large scale *in vitro* micropropagation of an ornamental plant, *Oxalis triangularis* A.st.-Hil, for commercial application. *Acta Horticulturae* 1339, 245–255.

Rittirat, S., Klaocheed, S., Kalawong, S. and Thammasiri, K. (2023) Adventitious Shoot Organogenesis from Shoot Tip Explants of an Ornamental Aquatic Plant *Anubias barteri* Var. *nana*. Proceedings of the 9th International Agriculture Congress (9th IAC). Çukurova University, Turkey, pp. 20–26.

Rittirat, S., Klaocheed, S., Kalawong, S. and Thammasiri, K. (2024) Asymbiotic Seed Germination and Plantlet Development of *Spathoglottis plicata* Blume, a Handsome Terrestrial Orchid. Proceedings of the 23rd World Orchid Conference. Tainan, Taiwan, pp. 248–251.

Saensouk, P., Muangsan, N., Saensouk, S. and Sirinajun, P. (2016) *In vitro* propagation of *Kaempferia marginata* Carey ex Roscoe, a native plant species to Thailand. *Journal of Animal and Plant Sciences* 26, 1405–1410. http://www.thejaps.org.pk/docs/v-26-05/30.pdf (accessed 4 May 2025).

Saensouk, P., Saensouk, S. and Pimmuen, P. (2018) *In vitro* propagation of *Globba schomburgkii* Hook. f. via bulbil explants. *Walailak Journal of Science and Technology* 15, 701–710. DOI: 10.48048/wjst.2018.3507

Saliwan, S., Saensouk, S. and P. Saensouk. (2022) *In vitro* propagation of *Kaempferia koratensis* Picheans., an endemic plant of Thailand. *Koch Cha Sarn Journal of Science* 44, 1–13. https://li01.tci-thaijo.org/index.php/kochasarn/article/view/253651 (accessed 4 May 2025).

Samala, S., Taweerodjanakarn, S., Thipwong, J., Thammasiri, K. and Chareonsap, P. (2020) Effects of BA and NAA on shoot multiplication of *Grammatophyllum speciosum* Blume *in vitro*. *Acta Horticulturae* 1298, 273–280.

Sangdum, S., Thammasiri, K., Chuenboonngarm, N. and Sopalun, K. (2017) Micropropagation of *Dendrobium cruentum* Rchb. f., a rare Thai orchid species. *Acta Horticulturae* 1167, 69–74.

Schenk, R.U. and Hildebrandt, A.C. (1972) Medium and techniques for induction and growth of monocotyledonous and dicotyledonous plant cell cultures. *Canadian Journal of Botany* 50, 199–204.

Siringama, T. and Pongprayoon, W. (2022) Influence of 6-benzylaminopurine (BAP) and indole-3-butyric acid (IBA) on shoot and root inductions of *Curcuma comosa* Roxb. through *vitro* culture. *Health Science and Technology Reviews* 15(2), 99–110. https://li01.tci-thaijo.org/index.php/journalup/article/view/252602 (accessed 4 May 2025).

Skoog, F. and Miller, C.O. (1957) Chemical regulation of growth and organ formation in plant tissue culture *in vitro*. *Symposium of the Society of Experimental Biology* 11, 118–131.

Sopalun, K., Thammasiri, K. and Ishikawa, K. (2010) Micropropagation of the Thai orchid *Grammatophyllum speciosum* blume. *Plant Cell Tissue and Organ Culture* 101, 143–150 (2010). DOI: 10.1007/s11240-010-9671-2

Steward, F.C., Mapes, M.O. and Mears, K. (1958). Growth and organized development of cultured cells. II. Organization in cultures grown from freely suspended cells. *American Journal of Botany* 705–708.

Thammasiri, K. (1997) *Orchids in Thailand: A Success Story*. APAARI, FAO, and Angkor Publishers, New Delhi, India.

Thammasiri, K., Jitsopakul, N. and Prasongsom, S. (2020) Micropropagation of Some Orchids and the Use of Cryopreservation. In: Merillon, J.M. and Kodja, H. (eds.) *Orchids Phytochemistry, Biology and Horticulture*. Springer Nature, Cham, Switzerland, pp. 225–260.

Thomas, T.D. (2008). The role of activated charcoal in plant tissue culture. *Biotechnology Advances* 26, 618–631. DOI: 10.1016/j.biotechadv.2008.08.003

Tilarux, P., Suwanposri, A., Tachai, S., Sasivatchutikool, P. and Boonmee, Y. (2022) Effect of plant hormones on growth of Wan Khan Mak (*Aglaonema tenuipes* Engl.) *in vitro*. *Journal of Science and Technology Mahasarakham University* 41, 227–235.

Vacin, E.F. and Went, F.W. (1949) Some pH changes in nutrient solutions. *Botanical Gazette* 110, 605–613.

Wu, I.F., Chen, J.T. and Chang, W.C. (2004) Effects of auxins and cytokinins on embryo formation from callus of *Oncidium* 'Gower Ramsey'. *Plant Cell and Tissue Organ Culture* 77, 107–109.

6 Molecular Insights into Developmental Pathways of Micropropagation in *Phalaenopsis* Orchids

Su-Chiung Fang*
Biotechnology Center in Southern Taiwan; Agricultural Biotechnology Research Center, Academia Sinica, Tainan, Taiwan

Abstract

The distinct floral morphology and physiological characteristics of orchids fascinate many evolutionary and plant biologists, and the beauty of these plants draws enormous enthusiasm from the public. To fulfil the demand for orchid plants among the general public, micropropagation techniques are essential for mass production of clonal plantlets in the orchid industry. Explants such as leaves, inflorescences, stems, and roots are commonly used for micropropagation. The ability of tissue to regenerate is a very important consideration for micropropagation as plantlets are usually regenerated from totipotent cells derived from dissected meristems, somatic tissue-induced protocorm-like bodies, and/or callus tissues. However, the molecular mechanisms underlying each different type of regeneration are not clear. Also, the developmental pathway undertaken for the type of regeneration used for orchid micropropagation is debatable. This chapter summarizes recent insights into the molecular mechanisms of micropropagation and discusses how this knowledge can benefit orchid breeding programs.

Keywords: *Phalaenopsis aphrodite*, orchid, protocorm-like body, organogenesis, somatic embryogenesis, shoot regeneration

6.1 Introduction

With more than 28,000 species (Christenhusz and Byng, 2016), Orchidaceae constitutes a large family in the plant kingdom. This family has adapted to a variety of diverse environments and is widely distributed. Evolutionarily speaking, orchids have always been a fascinating topic. Darwin himself was intrigued by orchids, with their diversity and the many ways by which orchid flowers and pollinators have co-evolved. While evolutionary biologists never lose interest in understanding how such diversity has evolved, the general public and even royalty also continually appreciate the beauty of the distinct and diverse floral structures found in orchids. The desire to study and/or cultivate orchids has spawned long-term tissue culture efforts to propagate orchid plantlets (Arditti, 2009; Arditti and Ernst, 1993; Yam and Arditti, 2017, 2018).

Micropropagation is defined as "an aseptic procedure for the asexual production of plantlets from organs, tissues, and cells bypassing the sexual process or other means of asexual propagation"

*Corresponding author: scfang@gate.sinica.edu.tw

 Biotechnology of Orchids and Other Ornamentals (Eds P. Bolaños-Villegas and F-C. Chen)
DOI: 10.1079/9781800627895.0006

(Hartmann and Kester, 1963; Krikorian, 1982). Micropropagation is important to propagate clonal plantlets in the orchid industry. However, this long process becomes a major bottleneck in achieving timely production of new cultivars to meet market demand. Devoted effort from scientific and industrial sectors has improved micropropagation techniques over the years. So far, protocorm-like bodies (PLBs) and direct shoot regeneration are the most common ways to propagate orchid plantlets. However, the developmental programs undertaken for each propagation means and the regulatory pathways governing the distinct regeneration processes remain uncertain. One obvious reason for this is that orchids are non-model systems and have long-life cycles (~3 years for *Phalaenopsis* orchid for example). Additionally, unlike model systems, orchids lack general genome resources and are not equipped with a modern molecular toolkit.

The recent establishment of the orchid genome and transcriptome databases (Ai *et al.*, 2021; Cai *et al.*, 2015; Chao *et al.*, 2018, 2017; Fang *et al.*, 2016; Li *et al.*, 2022; Rao *et al.*, 2014; Zhang *et al.*, 2017) provide unprecedented opportunities to study orchid biology. Indeed, work on comparative orchid genomics has suggested that expansions and contractions of gene families and acquisition of new gene families contribute to the diverse developmental programs and specialized innovations seen during orchid evolution (Zhang *et al.*, 2017). Genome-wide gene expression profiling studies suggest that duplication of regulators of organ identity such as the MADS box genes (an acronym for homeotic genes ***M****INICHROMOSOME MAINTENANCE 1* from budding yeast, ***A****GAMOUS* from *Arabidopsis thaliana*, ***D****EFICIENS* from snapdragon, and ***S****ERUM RESPONSE FACTOR* from humans) and the modified ABCDE model are important for specialized orchid floral organs (Ai *et al.*, 2021; Cai *et al.*, 2015; Su *et al.*, 2013). Additionally, the study of comparative orchid chloroplast genomes yields new insight into diversification of new orchid species (Givnish *et al.*, 2015). Moreover, a comprehensive transcriptome database of *Phalaenopsis* meristematic tissues has helped to resolve a long-time debate on the developmental origin of the specialized structure of the protocorm-like-body (Fang *et al.*, 2016). In addition to genome and transcriptome sources, progress has been made in developing molecular tools for manipulation and analysis of cellular processes in orchids (Chen and Fang, 2016; Hsing *et al.*, 2016; Liau *et al.*, 2003; Lin *et al.*, 2018; Lu *et al.*, 2012; Ren *et al.*, 2021). Therefore, it is now feasible to functionally characterize orchid genes and dissect the signal transduction pathways and molecular networks governing the evolutionary innovations of orchid species. With the advancement of molecular studies, herein I review recent progress in orchid micropropagation with a focus on molecular and developmental pathways governing PLBs, somatic embryogenesis, and direct shoot regeneration.

6.2 Molecular Identity of PLBs

The terminology "protocorm-like-body (PLB)" was first introduced by Professor Georges Morel to describe the "small flat bulblet looking exactly like a protocorm" (Arditti, 2022; Yam and Arditti, 2009). The protocorm is a peculiar structure emerging from a germinated orchid seed. Because orchid embryogenesis produces embryos that lack organogenesis and cotyledons, the protocorm does not have cotyledons upon germination and has only one anterior meristem where the new leaf is initiated first and roots emerge much later after germination (Nishimura, 1981; Yam *et al.*, 2002). Unlike the protocorm, which is derived from sexual reproduction, the PLB is triggered directly from explants and/or indirectly from explant-induced calluses (Chugh *et al.*, 2009; Jones and Tisserat, 1990; Wimber, 1963). Similar to the protocorm, the leaf primordium is initiated from the shoot pole of the PLB (Hong *et al.*, 2008) and each PLB grows into one plantlet. Because multiple PLBs can be induced from somatic tissues, PLB induction and propagation are often used to produce clonal plantlets in the orchid industry (Arditti, 2009; Yam and Arditti, 2017, 2018).

PLBs have been referred to as somatic embryos in the past because they resemble the embryonic structures emerging from embryogenesis (Chen and Chang, 2006; Gow *et al.*, 2009, 2010; Tokuhara and Mii, 2001, 2003). However, the molecular mechanism governing PLB development remains unclear. Previously, we used comparative transcriptome analysis to profile meristematic tissues including protocorms, PLBs, developing ovaries, floral stalks, stalk buds, and young leaves of *Phalaenopsis aphrodite* to identify the characteristic cellular activities in each tissue type (Fang *et al.*, 2016). Our findings showed that molecular components required for ovule and embryo development are evolutionally conserved in *P. aphrodite.* This study also provided crucial evidence indicating that the protocorm and PLB, which share similar transcriptome signatures, have very different cellular activities from those of zygotic embryos (Fig. 6.1A,B). Consistent with the transcriptomic analysis, the evolutionarily conserved embryonic genes such as *PaLEC1* and *PaBBM* that are expressed in developing orchid embryos (Fig. 6.1C) fail to be expressed in the developing PLBs (Fig. 6.1D), arguing again that PLBs are not somatic embryos. From these findings, we conclude that PLB is not a somatic embryo.

Importantly during the course of our study, *SHOOT MERISTEMLESS* (*PaSTM*), a class I KNOTTED-LIKE HOMEOBOX transcription factor, was identified as a PLB-associated gene (Fang, 2021; Fang *et al.*, 2016). The expression of gene *PaSTM* is tightly associated with PLB initiation and gradual decline but is maintained at moderate levels as development proceeds (Fig. 6.2A,B). Even though *PaSTM* was expressed in protocorms, its level was relatively low and restricted to the anterior end where shoots initiate (Fig. 6.2B). This expression pattern is consistent with the meristematic function of *STM* during organogenic shoot formation in *Arabidopsis*, maize, rice, tomato, *Kalanchoë daigremontiana*, *Agave tequilana*, and *Cuscuta pentagona* (dodder) (Abraham-Juárez *et al.*, 2010; Alakonya *et al.*, 2012; Bolduc and Hake, 2009; Chen *et al.*, 1997; Clark *et al.*, 1996; Garcês *et al.*, 2007; Hareven *et al.*, 1996; Tsuda *et al.*, 2011). *STM* and its homologs have been shown to regulate shoot meristem formation and maintain meristem activity in *Arabidopsis* and rice (Clark *et al.*, 1996; Long *et al.*, 1996; Tsuda *et al.*, 2011). Moreover, gain-of-function functional studies demonstrated that class I KNOX genes are sufficient to induce formation of tissue organization centers, which subsequently develop into ectopic shoot or organ-related structures (Atta *et al.*, 2009; Golz *et al.*, 2002; Gordon *et al.*, 2007; Lenhard *et al.*, 2002; Müller *et al.*, 1995; Sinha *et al.*, 1993; Williams-Carrier *et al.*, 1997). Similar to the maize *STM* homolog *KNOTTED1*, which directly regulates a *GA2 oxidase* to keep GA at a low level in the shoot apical meristem (Bolduc and Hake, 2009), expression of *PaSTM* in PLBs is positively correlated with its potential target *GIBBERELLIN 2-OXIDASE1* (*PaGA2OX1*) and a shoot regeneration marker *PaERF1* (the *Arabidopsis RAP2.6L* ERF/AP2 TF homolog; Fig. 6.2A). This suggests that PLBs have shoot generation capacity (Che *et al.*, 2006; Fang *et al.*, 2016). Based on this evidence, we hypothesized that the PLB is a modified version of the shoot, and *PaSTM* is important for PLB development.

Previously, we showed that *PaSTM* restored the shoot apical meristem defect of the *Arabidopsis stm-2* mutant (Fang *et al.*, 2022), supporting the meristematic function of *PaSTM*. To investigate the functional role of *PaSTM* in PLB development, we generated transgenic orchid plants overexpressing *PaSTM* and showed that *PaSTM* is sufficient to enhance PLB generation in three independent T0 transgenic lines (Fig. 6.3A). To dissect the regulatory pathway of PLB development, a comparative transcriptomic approach was deployed. It is already known that despite sharing a similar regulatory network, the PLB and protocorm are not regulated in an identical manner (Fang *et al.*, 2016). We reasoned that genes involved in PLB development may be specifically upregulated in PLBs and sought to identify these genes. After comparing the transcriptomic dynamics of the categorized PLBs and protocorms, we identified 235 PLB-specific genes (Fang *et al.*, 2022). Among them, 14 were transcription factors and six of them were induced by *PaSTM*. Interestingly, a group of S11 MYB transcription factors, *PaMYB13*, *PaMYB14*, and *PaMYB17*, are coordinately activated by *PaSTM* (Fig. 6.3B,C). In addition, the promoters of S11 MYB transcription factors, including the ones from *Arabidopsis*, maize, and rice share the TTGACT motif (Fang *et al.*, 2022), which carries the conserved KNOTTED1-binding core TGAC (Bolduc *et al.*, 2014; Smith *et al.*, 2002). We used electrophoretic mobility shift assay to demonstrate that this TTGACT motif is required for binding of the PaSTM protein (Fang *et al.*, 2022). Together, these findings suggest that *PaMYB13*, *PaMYB14*, and *PaMYB17* are the direct targets of *PaSTM* in *Phalaenopsis* orchids. Supporting this notion, ectopic expression of *PaSTM* led to an increase in levels of the

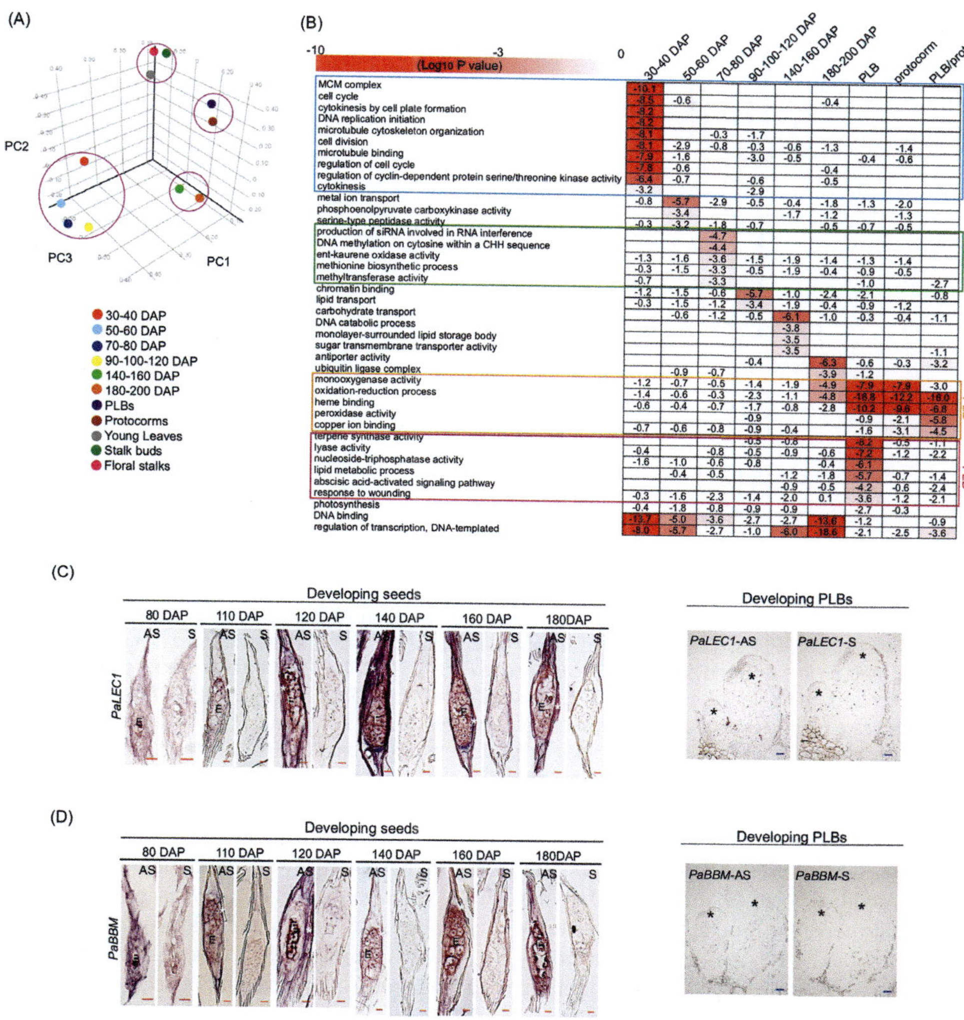

Fig. 6.1. Distinct molecular signatures are associated with different reproductive tissues. A) Principal component analysis of developing ovaries at 30–40 days after pollination (DAP), 50–60 DAP, 70–80 DAP, 90–100–120 DAP, 140–160 DAP, and 180–200 DAP, as well as PLBs, protocorms, young leaves, stalk buds, and floral stalks. B) Heat map showing the statistical signficance (*p* value) of enriched Gene Ontology (GO) terms for stage-specific mRNAs. Empty cells indicate lack of association. C) *In situ* hybridization with an anti-sense *PaLEC1* probe (*PaLEC1*-AS) on longitudinal sections through the center of the developing seeds at different DAP and the developing PLBs. Sense probe of *PaLEC1* (*PaLEC1*-S) was used as a negative control. D) *In situ* hybridization with an anti-sense *PaBBM* probe (*PaBBM*-AS) on longitudinal sections through the center of the developing seeds at different DAP and the developing PLBs. The sense probe of *PaBBM* (*PaBBM*-S) was used as a negative control. DAP, days after pollination. E, embryo; white asterisk (*), seed coat; black asterisk (*), individual PLB. Red scale bar = 20 μm. Blue scale bar = 200 μm. Modified from Fang *et al.* (2016). Copyright American Society of Plant Biologists (CC BY).

Arabidopsis S11 *MYB* transcription factors, *MYB102* and *MYB74*, but not *MYB41*, which happens to lack the TTGACT consensus, arguing the functional importance of the TTGACT element for *PaSTM*-dependent regulation.

Unlike organogenesis commonly seen in other plant species, where organs such as shoots and roots are directly generated (Duclercq *et al.*, 2011), the PLB appears to act as an intermediate structure before the emergence of

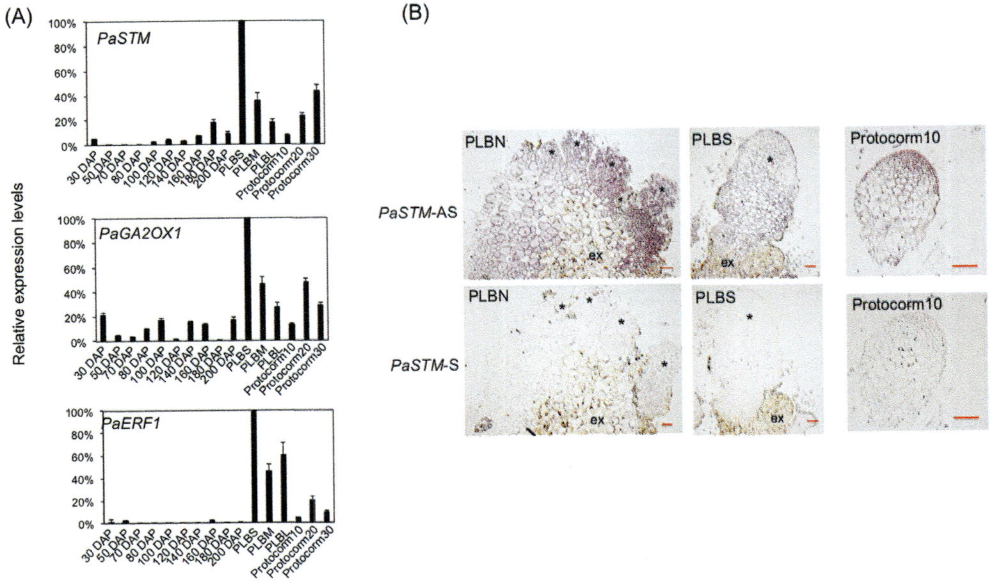

Fig. 6.2. Expression of *PaSTM* mRNA is associated with PLB initiation. A) Expression patterns of the indicated genes in the developing PLBs, protocorms, and interior ovary tissues collected at the indicated day after pollination (DAP). B) *In situ* hybridization with an anti-sense *PaSTM* (*PaSTM*-AS) probe on longitudinal sections through the center of newly emerging PLBs (PLBN) and small-sized PLBs (PLBS). Sense *PaSTM* (*PaSTM*-S) probe is used as a negative control. PLBN, newly emerging PLBs; PLBS, small-sized PLBs; PLBM, medium-sized PLBs; PLBL, large-sized PLBs; Protocorm10, 10-day-old protocorms; Protocorm20, 20-day-old protocorms; Protocorm30, 30-day-old protocorms; *, individual PLB; ex, explant tissues from which PLBs are derived. Scale bar = 200 μm. Modified from Fang *et al.* (2016). Copyright American Society of Plant Biologists (CC BY).

new shoots. We therefore hypothesized that the PLB is a modified form of an organ that has shoot apical meristem function and eventually generates a shoot. This hypothesis was tested using the *de novo* organogenesis (shoot induction) assay established for *Arabidopsis* (Che *et al.*, 2006; Ikeuchi *et al.*, 2017). Indeed, overexpression of *PaSTM* enhanced shoot regeneration ability by increasing the number of emerging leaves on hypocotyl explants (Fig. 6.3D), supporting the role of *PaSTM* to enhance shoot-related organogenesis (Fang *et I al.*, 2022). Moreover, overexpression of *PaMYB13*, *PaMYB14*, or *PaMYB17*, the direct targets of PaSTM, also improved shoot regeneration ability, reaffirming that the PaSTM-S11 MYB regulatory module is involved in establishing shoot or shoot-related organ development (Fig. 6.3D). It is important to note that ectopic PaSTM only induced PLBs but did not activate direct shoot development, suggesting a distinct role for PaSTM in *Phalaenopsis* orchids. In summary, these results support an evolutionarily conserved role for PaSTM and also suggest the co-option of the STM-S11 MYB regulatory pathway in regulating PLB (shoot-related) organogenesis in *Phalaenopsis* orchids and potentially in other orchid species.

Even though PLBs can be induced from explants of leaves, nodes, shoot tips, stems, root tips, and protocorms (Paek *et al.*, 2011; Teixeira da Silva, 2013; Yam and Arditti, 2017), the response of an explant to *in vitro* conditions varies widely between families, genera, species, hybrids, and genotypes, and even environmental conditions (Yam and Arditti, 2017). Considering the ability of *PaSTM* to induce PLB regeneration, *PaSTM* may be used as a molecular marker to assist screening for progeny that enable PLB-based micropropagation in the orchid breeding program.

6.3 Diverse Somatic Embryonic Structures Induced in *Phalaenopsis* Orchids

If the PLB were not a somatic embryo, we may wonder what the somatic embryo of *Phalaenopsis*

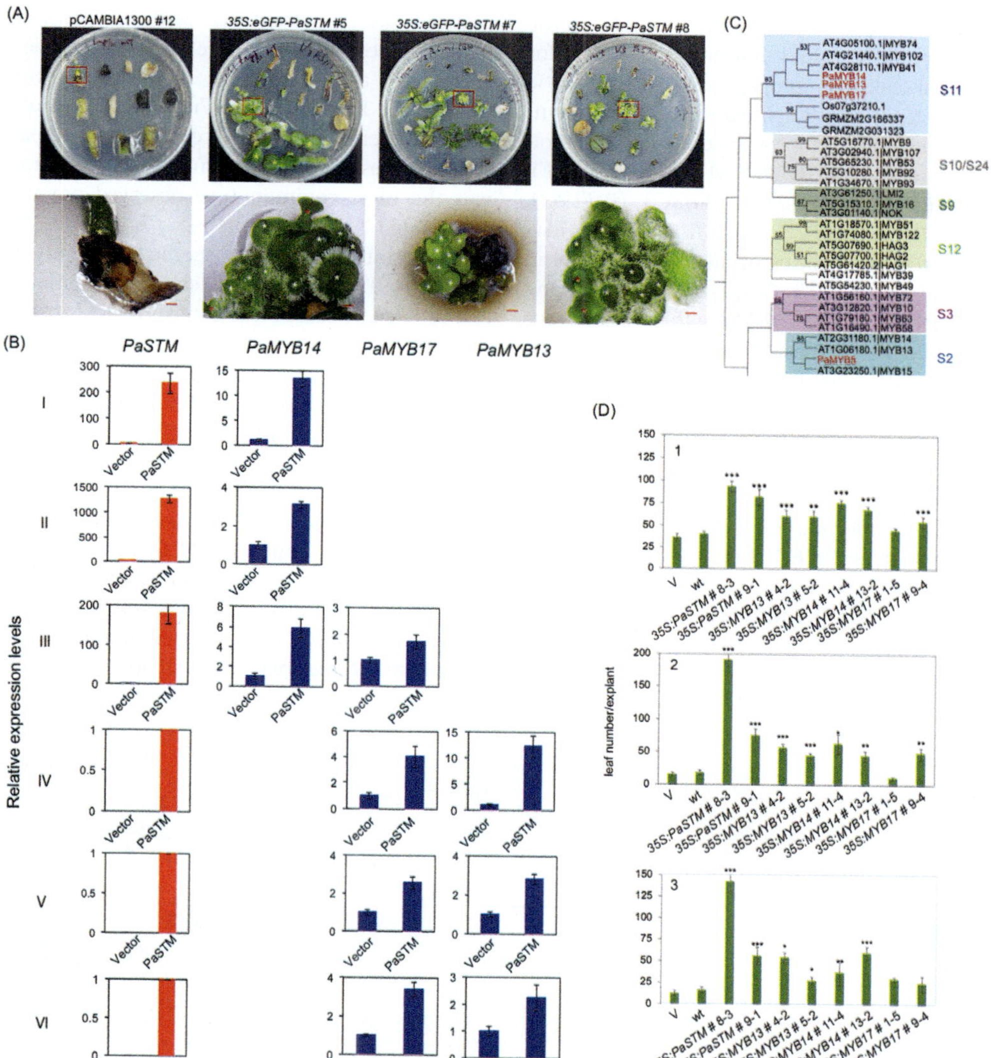

Fig. 6.3. Overexpression of *PaSTM-eGFP* in *Phalaenopsis* orchids enhances PLB generation. A) PLB induction from explants of T0 *35S:PaSTM-eGFP* transgenic plantlets. Scale bar, 1 mm. The tissues in the red rectangle of the top panels were enlarged in the bottom panel. Notice that new shoots (marked by yellow asterisks) were able to grow from explants containing the axillary meristem. White asterisk (*), individual PLB; red asterisk (*), PLB with a developing leaf. B) Effect of *PaSTM* overexpression on the S11 *MYB* transcription factors *PaMYB13*, *PaMYB14*, and *PaMYB17*. *PaSTM* was overexpressed transiently by agroinfiltration in orchid leaves. The second or third leaf of orchid plants was used for agroinfiltration. The agroinfiltration experiment was repeated six times. Upregulation of transcription factors (TFs) was confirmed by at least three independent experiments. Roman numbers represent independent experiments. Data are from three technical replicates and the error bars are presented as standard error of the mean. C) A snapshot of the phylogenetic relationship of MYB transcription factors of *Arabidopsis thaliana* (AT) and selected members from *Phalaenopsis aphrodite* (Pa), *Oryza sativa* (Os), and *Zea mays* (GRMZM). D) Quantification of shoot induction assays. The number of leaves ± standard error of the mean (SEM) regenerated from hypocotyl explants was counted after 9 or 10 days on the Sulfide-Indole Motility (SIM) medium. Each genotype had at least 8 explants and the experiment was repeated 3 times. All statistical significance values (*p* values) correspond to the Student's *t* test. *, $p < 0.05$; **, $p < 0.01$; ***, $p < 0.01$. All three experiments (labelled as 1, 2, and 3) are shown. Modified from Fang *et al.* (2022). Copyright American Society of Plant Biologists. Reproduced with permission from Oxford University Press.

orchids would like and whether orchid somatic embryogenesis would have the potential for industrial-scale micropropagation. During zygotic embryogenesis of *P. aphrodite*, many evolutionarily conserved embryo-specific transcription factors such as *BABY BOOM* (*PaBBM*), *LEAFY COTYLEDON 1* (*PaLEC1*), *FUSCA* (*PaFUS3*), *PaABI3L1*, *PaABI3L2*, and *WRINKLED1* (*PaWRI1*) were found to be associated with developing embryos (Fang, 2021; Fang *et al.*, 2016). Among them, *PaLEC1* and *PaBBM* are the master regulators of embryogenesis (Braybrook and Harada, 2008) whose expression is not only associated with developing orchid embryos, but is also crucial in activating the somatic embryonic program (Fang *et al.*, 2016). To gain comprehensive understanding of the somatic embryogenesis program of *Phalaenopsis* orchids, the embryonic-specific gene *PaLEC1* fused with the *GFP* reporter was allowed to overexpress in *P. equestris* (Chen *et al.*, 2019). Unlike wild-type protocorms that grow into seedlings (Fig. 6.4A), overexpressing the PaLEC1-GFP protein in transgenic plants induced diverse structures with categorized morphologies similar to the callus, multiple-meristem, and dorsal-fin structures from germinated T1 protocorms (Fig. 6.4B). The multiple-meristem type (tissues containing multiple shoot apexes) of somatic embryonic structure was the most common type observed among germinated T1 protocorms (48.1% and 56.3% from two independent transgenic plants (Chen *et al.*, 2019). The embryonic identity of somatic embryonic structures was confirmed by expression of the embryonic-specific transcription factors *FUSCA3* (*PeFUS3*) and *ABSCISIC ACID INSENSITIVE3* (*PeABI3*), and seed storage proteins *7S GLOBULIN* and *OLEOSIN* (Chen *et al.*, 2019). Furthermore, we used chromatin immunoprecipitation to show that PaLEC1-GFP fusion protein directly binds to the embryonic genes such as *Pe7S-1* and *PeFUS3* to activate the somatic embryonic program in *P. equestris* (Chen *et al.*, 2019). Regardless of the types of embryonic structures, *PaLEC1-GFP*-induced somatic embryonic structures are pluripotent and capable of

Fig. 6.4. Developmental processes of T1 *35:PaLEC1-GFP* #1 transgenic plants. A) Germinated protocorms of wild-type plants developed into young seedlings at 1, 2, 3, and 4 months after germination. B) Germinated protocorms of *35:PaLEC1-GFP* #1 transgenic plants developed into callus-type, multiple-meristem type, and dorsal fin-like structures. MAS, month after sowing; DAS, day after sowing. Black scale bar = 100 μm. Red scale bar = 500 μm. Blue scale bar = 1 mm. Modified from Chen *et al.* (2019). Copyright Frontiers in Plant Science (CC BY).

generating new shoots. It is important to note that the *PaLEC1*-induced somatic embryonic structures are different from direct shoot regeneration during which leaf primordia emerge directly (Košir *et al.*, 2004; Wu and Chen, 2008b; Yam and Arditti, 2017). Importantly, *PaLEC1*-induced somatic embryonic tissues are morphologically different from those of PLBs, supporting our conclusion that PLB is not a somatic embryo. This study strengthened our conclusion about the developmental identity of the PLB and suggests that somatic embryogenesis is a useful strategy for orchid micropropagation.

6.4 Direct Shoot Regeneration

In addition to the PLB, the most common way to propagate orchid plantlets is to induce *de novo* shoots directly from nodal buds of floral stalks (Fig. 6.5). Based on the literature and its accounts of micropropagation (Arditti, 2022; Yam and Arditti, 2017, 2018), Dr Gavino Rotor, inspired by a lecture given by Professor Lewis Knudson, developed the first stalk bud-based shoot regeneration protocol for *Phalaenopsis* orchids (Rotor, 1949). Since then, inflorescence explants of other orchids such as *Aranda* (Goh and Wong, 1990), *Dendrobium* (Singh and Sagawa, 1972), *Doritaenopsis* (Ernst, 1994; Tokuhara and Mii, 1993; Wu and Chen, 2008a), *Oncidium* (Chen and Chang, 2000), and *Phalaenopsis* (Chen and Piluek, 1995; Tanaka *et al.*, 1988; Tanaka and Sakanishi, 1978; Tokuhara and Mii, 1993, 1998, 2001) have been applied for micropropagation.

Direct shoot regeneration from floral-stalk nodes (Bouriquet *et al.*, 1982) is inefficient for generating clonal plantlets directly because only one or a few shoots can be induced from one floral nodal bud (Wu and Chen, 2008a,b). However, in contrast to PLB-based clonal propagation that has relatively high somaclonal variation, (Chen and Chen, 2007; Chen *et al.*, 2008, 1998), nodal bud-based shooting, which has relatively low somaclonal variation, is often preferred in the orchid industry. Even though meristem-based shoot regeneration is common in nature, it remains unclear what molecular mechanism was undertaken to control *de novo* shoot assembly in orchids. Recently, we developed an efficient shoot regeneration technology (Fang and Chang, 2023), that boosted the efficiency of generation of clonal plantlets dramatically. This shoot regeneration platform has not only been demonstrated to enable propagation of multiple commercialized cultivars, but it also provides us with a unique opportunity to probe the molecular mechanism of *de novo* shoot assembly of *Phalaenopsis* orchids. So far, our preliminary data suggests that the developmental pathway of direct shoot regeneration is different from that of PLB development (data not shown). It would be interesting to investigate how orchid somatic cells are reprogrammed to activate shoot development.

6.5 Conclusion and Perspectives

Micropropagation is important for generating clonal orchid plantlets in the orchid industry.

Fig. 6.5. Three independent developmental pathways of micropropagation in *Phalaenopsis* orchids: PLBs, somatic embryogenesis, and direct shoot regeneration. Modified from Chen *et al.* (2019). Copyright Frontiers in Plant Science.

Here, I summarized studies related to three independent micropropagation pathways: PLB, somatic embryogenesis, and direct shoot regeneration (Fig. 6.5). Our genome-wide comparative transcriptomic and molecular genetic studies resolved a long-time debate on the developmental origin of PLB through which we concluded that PLB is not a somatic embryo. In addition, our results showed that overexpression of *Phalaenopsis aphrodite STM* (*PaSTM*) greatly enhances PLB regeneration from vegetative tissue-based explants of *Phalaenopsis* orchids, confirming its regulatory role in PLB development. The class S11 MYB transcription factors were identified as direct targets downstream of *PaSTM*. Together with *PaSTM*, the class S11 *MYB* transcription factors *PaMYB13*, *PaMYB14*, and *PaMYB17*, enhanced *de novo* shoot regeneration in *Arabidopsis*, indicating the active role of the *PaSTM-S11* and *PaMYB* module in organogenesis. Hence, we propose that PLB is derived from organogenesis that is similar but not identical to shoot development (Fang *et al.*, 2022). In addition, we used the embryonic inducer *PaLEC1* to show the diverse somatic embryonic structures of *Phalaenopsis* orchids for the first time and demonstrated their ability to propagate orchid plantlets (Chen *et al.*, 2019). This study also strengthens our conclusion that PLB is not a somatic embryo. Lastly, we established a direct shoot regeneration platform that enables us to investigate the molecular and cellular dynamics of *de novo* shoot assembly in *Phalaenopsis* orchids. Further study to investigate how somatic cells switch fate and transition to shoot development will enable fundamental understanding of shoot development in orchids. Importantly, the shoot "switches" that are identified may be useful markers for breeding programs in orchid biotechnology in Taiwan and elsewhere.

Acknowledgments

This study was supported by the Innovative Translational Agricultural Research Grants (to SCF); and in part by a grant (to SCF) from the Biotechnology Center in Southern Taiwan, Academia Sinica. I thank Jhun-Chen Chen, Pou-Yi Chang, Hsiang-Yin Lin, Dr. Chii-Gong Tong, Pei-Miao Chien, and Miao-Ju Wei, for their contribution to the described studies and Miranda Loney for English editing.

References

Abraham-Juárez, M.J., Martínez-Hernández, A., Leyva-González, M.A., Herrera-Estrella, L. and Simpson, J. (2010) Class I *KNOX* genes are associated with organogenesis during bulbil formation in *Agave tequilana*. *Journal of Experimental Botany* 61, 4055–4067.

Ai, Y., Li, Z., Sun, W.H., Chen, J., Zhang, D. *et al.* (2021) The *Cymbidium* genome reveals the evolution of unique morphological traits. *Horticulture Research* 8, 255.

Alakonya, A., Kumar, R., Koenig, D., Kimura, S., Townsley, B., Runo, S., Garces, H.M., Kang, J., Yanez, A., David-Schwartz, R., Machuka, J. and Sinha, N. (2012) Interspecific RNA interference of *SHOOT MERISTEMLESS*-like disrupts *Cuscuta pentagona* plant parasitism. *Plant Cell* 24, 3153–3166.

Arditti, J. (2009) *Micropropagation of Orchids*, 2nd edn. Wiley-Blackwell, Oxford, UK.

Arditti, J. (2022) Orchids- discoverers and firsts. *South African Journal of Botany* 146, 815–848.

Arditti, J. and Ernst, R. (1993) *Micropropagation of Orchids*. Wiley, New York/USA.

Atta, R., Laurens, L., Boucheron-Dubuisson, E., Guivarc'h, A., Carnero, E. *et al.* (2009) Pluripotency of *Arabidopsis* xylem pericycle underlies shoot regeneration from root and hypocotyl explants grown *in vitro*. *Plant Journal* 57, 626–644.

Bolduc, N. and Hake, S. (2009) The maize transcription factor *KNOTTED1* directly regulates the gibberellin catabolism gene ga2ox1. *Plant Cell* 21, 1647–1658.

Bolduc, N., Tyers, R.G., Freeling, M. and Hake, S. (2014) Unequal redundancy in maize knotted1 homeobox genes. *Plant Physiology* 164, 229–238.

Bouriquet, R., Broly, H. and Legrand, B. (1982) Clonal propagation of *Phalaenopsis* (Orchidaceae) by *in vitro* culture. In: Earle, E.D. and Demarly, Y. (eds) *Variability in Plants Regeneration from Tissue Culture*. Praeger Publishers, New York, pp. 35–46.

Braybrook, S.A. and Harada, J.J. (2008) LECs go crazy in embryo development. *Trends in Plant Science* 13, 624–630.

Cai, J., Liu, X., Vanneste, K., Proost, S., Tsai, W.C. *et al.* (2015) The genome sequence of the orchid *Phalaenopsis equestris*. *Nature Genetics* 47, 65–72.

Chao, Y.T., Yen, S.H., Yeh, J.H., Chen, W.C. and Shih, M.C. (2017) Orchidstra 2.0-a transcriptomics resource for the orchid family. *Plant Cell Physiology* 58, 1–13.

Chao, Y.T., Chen, W.C., Chen, C.Y., Ho, H.Y., Yeh, C.H. *et al.* (2018) Chromosome-level assembly, genetic and physical mapping of *Phalaenopsis aphrodite* genome provides new insights into species adaptation and resources for orchid breeding. *Plant Biotechnology Journal* 16, 2027–2041.

Che, P., Lall, S., Nettleton, D. and Howell, S.H. (2006) Gene expression programs during shoot, root, and callus development in *Arabidopsis* tissue culture. *Plant Physiology* 141, 620–637.

Chen, J.T. and Chang, W.C. (2000) Efficient plant regeneration through somatic embryogenesis from callus cultures of *Oncidium* (Orchidaceae). *Plant Science* 160, 87–93.

Chen, J. and Chang, W. (2006) Direct somatic embryogenesis and plant regeneration from leaf explants of *Phalaenopsis amabilis. Biologia Plantarum* 50, 169–173.

Chen, F.-C. and Chen, W.-H. (2007) Somaclonal variation in orchids. In: Chen, W.-H. and Chen, H.-H. (eds.) *Orchid Biotechnology*. World Scientific, Singapore, pp. 65–76.

Chen, J.C. and Fang, S.C. (2016) The long pollen tube journey and *in vitro* pollen germination of *Phalaenopsis* orchids. *Plant Reproduction* 29, 179–188.

Chen, Y. and Piluek, C. (1995) Effects of thidiazuron and N6-benzylaminopurine on shoot regeneration of *Phalaenopsis*. *Plant Growth Regulation* 16, 99–101.

Chen, J.J., Janssen, B.J., Williams, A. and Sinha, N. (1997) A gene fusion at a homeobox locus: alterations in leaf shape and implications for morphological evolution. *Plant Cell* 9, 1289–1304.

Chen, W.H., Chen, T.M., Fu, Y.M., Hsieh, R.M. and Chen, W.S. (1998) Studies on somaclonal variation in *Phalaenopsis*. *Plant Cell Reports* 18, 7–13.

Chen, F. -C., Yu, J. Y., Chen, P. Y. and Huang, Y. W. (2008). Somaclonal variation in orchids and the application of biotechnology. *Acta Horticulturae*, 766, 315–322.

Chen, J.C., Tong, C.G., Lin, H.Y. and Fang, S.C. (2019) Phalaenopsis *LEAFY COTYLEDON1*-induced somatic embryonic structures are morphologically distinct from protocorm-like bodies. *Frontiers in Plant Science* 10, 1594.

Christenhusz, M.J.M. and Byng, J.W. (2016) The number of known plants species in the world and its annual increase. *Phytotaxa* 261, 201–217.

Chugh, S., Guha, S. and Rao, I.U. (2009) Micropropagation of orchids: a review on the potential of different explants. *Scientia Horticulturae* 122, 507–520.

Clark, S.E., Jacobsen, S.E., Levin, J.Z. and Meyerowitz, E.M. (1996) The *CLAVATA* and *SHOOT MERISTEMLESS* loci competitively regulate meristem activity in *Arabidopsis. Development* 122, 1567–1575.

Duclercq, J., Sangwan-Norreel, B., Catterou, M. and Sangwan, R.S. (2011) *De novo* shoot organogenesis: from art to science. *Trends in Plant Science* 16, 597–606.

Ernst, R. (1994) Effects of thidiazuron on *in vitro* propagation of *Phalaenopsis* and *Doritaenopsis* (Orchidaceae). *Plant Cell and Tissue Organ Culture* 39, 273–275.

Fang, S.C. (2021) Comparative transcriptomics study provides new insights into the specialized reproductive developmental programs of *Phalaenopsis aphrodite*. In: Chen, W.H. and Chen, H.H. (eds.) *Orchid Biotechnology IV*. World Scientific, Singapore, pp. 173–205.

Fang, S.C., Chen, J.C. and Wei, M.J. (2016) Protocorms and protocorm-like bodies are molecularly distinct from zygotic embryonic tissues in *Phalaenopsis aphrodite*. *Plant Physiology* 171, 2682–2700.

Fang, S.C., Chen, J.C., Chang, P.Y. and Lin, H.Y. (2022) Co-option of the *SHOOT MERISTEMLESS* network regulates protocorm-like body development in *Phalaenopsis aphrodite*. *Plant Physiology* 190, 127–145.

Fang, S.C. and Chang, P.Y. (2023) Improvement of clonal micropropagation in Phalaenopsis orchids. Docket Number: 4T-1070516. Department of Intellectual Property and Technology Transfer, Taiwan. https://iptt.sinica.edu.tw/shares/650

Garcês, H.M., Champagne, C.E., Townsley, B.T., Park, S., Malho, R., Pedroso, M.C., Harada, J.J. and Sinha, N.R. (2007) Evolution of asexual reproduction in leaves of the genus *Kalanchoë*. *Proceedings of the National Academies of Science of the USA* 104, 15578–15583.

Givnish, T.J., Spalink, D., Ames, M., Lyon, S.P., Hunter, S.J. *et al.* (2015) Orchid phylogenomics and multiple drivers of their extraordinary diversification. *Proceedings of the Royal Society B: Biological Sciences* 282, 20151553.

Goh, C.J. and Wong, P.F. (1990) Micropropagation of the monopodial orchid hybrid *Aranda* 'Deborah' using inflorescence explants. *Scientia Horticulturae* 44, 315–321.

Golz, J.F., Keck, E.J. and Hudson, A. (2002) Spontaneous mutations in *KNOX* genes give rise to a novel floral structure in *Antirrhinum*. *Current Biology* 12, 515–522.

Gordon, S.P., Heisler, M.G., Reddy, G.V., Ohno, C., Das, P. and Meyerowitz, E.M. (2007) Pattern formation during de novo assembly of the *Arabidopsis* shoot meristem. *Development* 134, 3539–3548.

Gow, W.P., Chen, J.T. and Chang, W.C. (2009) Effects of genotype, light regime, explant position and orientation on direct somatic embryogenesis from leaf explants of *Phalaenopsis* orchids. *Acta Physiologiae Plantarum* 31, 363–369.

Gow, W.P., Chen, J.T. and Chang, W.C. (2010) Enhancement of direct somatic embryogenesis and plantlet growth from leaf explants of *Phalaenopsis* by adjusting culture period and explant length. *Acta Physiologiae Plantarum* 32, 621–627.

Hareven, D., Gutfinger, T., Parnis, A., Eshed, Y. and Lifschitz, E. (1996) The making of a compound leaf: genetic manipulation of leaf architecture in tomato. *Cell* 84, 735–744.

Hartmann, H.T. and Kester, D.E. (1963) *Plant Propagation: Principles and Practice*. Prentice-Hall Englewood Cliffs, New Jersey/USA.

Hong, P.I., Chen, J.T. and Chang, W.C. (2008) Plant regeneration via protocorm-like body formation and shoot multiplication from seed-derived callus of a *maudiae* type slipper orchid. *Acta Physiologiae Plantarum* 30, 755–759.

Hsing, H.X., Lin, Y.J., Tong, C.G., Li, M.J., Chen, Y.J. and Ko, S.S. (2016) Efficient and heritable transformation of *Phalaenopsis* orchids. *Botanical Studies* 57, 1–12.

Ikeuchi, M., Iwase, A., Rymen, B., Lambolez, A., Kojima, M., Takebayashi, Y., Heyman, J., Watanabe, S., Seo, M., De Veylder, L., Sakakibara, H. and Sugimoto, K. (2017) Wounding triggers callus formation via dynamic hormonal and transcriptional changes. *Plant Physiology* 175, 1158–1174.

Jones, D. and Tisserat, B. (1990) Clonal propagation of orchids. *Methods in Molecular Biology* 6, 181–191.

Košir, D., Škof, S. and Luthar, Z. (2004) Direct shoot regeneration from nodes of *Phalaenopsis* orchids. *Acta Agriculturae Slovenica* 83, 233–242.

Krikorian, A.D. (1982) Cloning higher plants from aseptically cultured tissues and cells. *Biological Reviews* 57, 151–218.

Lenhard, M., Jürgens, G. and Laux, T. (2002) The *WUSCHEL* and *SHOOTMERISTEMLESS* genes fulfil complementary roles in Arabidopsis shoot meristem regulation. *Development* 129, 3195–3206.

Li, M.H., Liu, K.W., Li, Z., Lu, H.C., Ye, Q.L. *et al.* (2022) Genomes of leafy and leafless *Platanthera* orchids illuminate the evolution of mycoheterotrophy. *Nature Plants* 8, 373–388.

Liau, C.H., You, S.J., Prasad, V., Hsiao, H.H., Lu, J.C., Yang, N.S. and Chan, M.T. (2003) Agrobacterium tumefaciens-mediated transformation of an *Oncidium* orchid. *Plant Cell Reproduction* 21, 993–998.

Lin, H.Y., Chen, J.C. and Fang, S.C. (2018) A protoplast transient expression system to enable molecular, cellular, and functional studies in *Phalaenopsis* orchids. *Frontiers in Plant Science* 9, 843.

Long, J.A., Moan, E.I., Medford, J.I. and Barton, M.K. (1996) A member of the *KNOTTED* class of homeodomain proteins encoded by the STM gene of *Arabidopsis*. *Nature* 379, 66–69.

Lu, H.C., Hsieh, M.H., Chen, C.E., Chen, H.H., Wang, H.I. and Yeh, H.H. (2012) A high-throughput virus-induced gene-silencing vector for screening transcription factors in virus-induced plant defense response in orchid. *Molecular Plant Microbe Interactions* 25, 738–746.

Müller, K.J., Romano, N., Gerstner, O., Garcia-Maroto, F., Pozzi, C., Salamini, F. and Rohde, W. (1995) The barley *Hooded* mutation caused by a duplication in a homeobox gene intron. *Nature* 374, 727–730.

Nishimura, G. (1981) Comparative morphology of *Cattleya* and *Phalaenopsis* (Orchidaceae) seedlings. *Botanical Gazette* 142, 360–365.

Paek, K.Y., Hahn, E.J. and Park, S.Y. (2011) Micropropagation of *Phalaenopsis* orchids via protocorms and protocorm-like bodies. *Methods in Molecular Biology* 710, 293–306.

Rao, X., Krom, N., Tang, Y., Widiez, T., Havkin-Frenkel, D., Belanger, F.C., Dixon, R.A. and Chen, F. (2014) A deep transcriptomic analysis of pod development in the vanilla orchid (*Vanilla planifolia*). *BMC Genomics* 15, 964.

Ren, R., Gao, J., Yin, D., Li, K., Lu, C., Ahmad, S., Wei, Y., Jin, J., Zhu, G. and Yang, F. (2021) Highly efficient leaf base protoplast isolation and transient expression systems for orchids and other important monocot crops. *Frontiers in Plant Science* 12, 626015.

Rotor, G. (1949) A method of vegetative propagation of *Phalaenopsis* species and hybrids. *American Orchid Society Bulletin* 18, 738–739.

Singh, H. and Sagawa, Y. (1972) Vegetative propagation of *Dendrobium* by flower stalk cuttings. *Hawaii Orchid Journal* 1(4), 19.

Sinha, N.R., Williams, R.E. and Hake, S. (1993) Overexpression of the maize homeo box gene, *KNOTTED-1*, causes a switch from determinate to indeterminate cell fates. *Genes and Development* 7, 787–795.

Smith, H.M., Boschke, I. and Hake, S. (2002) Selective interaction of plant homeodomain proteins mediates high DNA-binding affinity. *Proceedings of the National Academies of Sciences of the USA* 99, 9579–9584.

Su, C.L., Chen, W.C., Lee, A.Y., Chen, C.Y., Chang, Y.C., Chao, Y.T. and Shih, M.C. (2013) A modified ABCDE model of flowering in orchids based on gene expression profiling studies of the moth orchid *Phalaenopsis aphrodite*. *PLoS One* 8: e80462.

Tanaka, M., Kumura, M. and Goi, M. (1988) Optimal conditions for shoot production from *Phalaenopsis* flower-stalk cuttings cultured *in vitro*. *Scientia Horticulturae* 35, 117–126.

Tanaka, M. and Sakanishi, Y. (1978) Factors affecting the growth of *in vitro* cultured lateral buds from *Phalaenopsis* flower stalks. *Scientia Horticulturae* 8, 169–178.

Teixeira da Silva, J.A. (2013) The role of thin cell layers in regeneration and transformation in orchids. *Plant Cell and Tissue Organ Culture* 113, 149–161.

Tokuhara, K. and Mii, M. (1993) Micropropagation of *Phalaenopsis* and *Doritaenopsis* by culturing shoot tips of flower stalk buds. *Plant Cell Reproduction* 13, 7–11.

Tokuhara, K. and Mii, M. (1998) Somaclonal variations in flower and inflorescence axis in micropropagated plants through flower stalk bud culture of *Phalaenopsis* and *Doritaenopsis. Plant Biotechnology* 15, 23–28.

Tokuhara, K. and Mii, M. (2001) Induction of embryogenic callus and cell suspension culture from shoot tips excised from flower stalk buds of *Phalaenopsis* (Orchidaceae). *In Vitro Cell Development Biology - Plants* 37, 457-461.

Tokuhara, K. and Mii, M. (2003) Highly-efficient somatic embryogenesis from cell suspension cultures of *Phalaenopsis* orchids by adjusting carbohydrate sources. *In Vitro Cell Development Biology - Plants* 39, 635-639.

Tsuda, K., Ito, Y., Sato, Y. and Kurata, N. (2011) Positive autoregulation of a *KNOX* gene is essential for shoot apical meristem maintenance in rice. *Plant Cell* 23, 4368–4381.

Williams-Carrier, R.E., Lie, Y.S., Hake, S. and Lemaux, P.G. (1997) Ectopic expression of the maize *kn1* gene phenocopies the *Hooded* mutant of barley. *Development* 124, 3737–3745.

Wimber, D.E. (1963) Clonal multiplication of cymbidiums through tissue culture of the shoot meristem. *American Orchid Society Bulletin* 32. https://www.osti.gov/biblio/12491054

Wu, H.-H. and Chen, F.-C. (2008a) Effect of plant growth regulators on shoot multiplication from flower stalk nodal buds of *Phalaenopsis* and *Doritaenopsis*. *Journal Taiwan Society of Horticultural Science* 54, 151–159.

Wu, H.-H. and Chen, F.-C. (2008b) Induction of flower stalk nodal buds of *Phalaenopsis* and *Doritaenopsis* and effect of wounding on shoot multiplication. *Journal of the Taiwan Society of Horticultural Science* 54, 67–74.

Yam, T.W. and Arditti, J. (2009) History of orchid propagation: a mirror of the history of biotechnology. *Plant Biotechnology Reproduction* 3, 1–56.

Yam, T.W. and Arditti, J. (2017) *Micropropagation of Orchids*, 3rd edn. John Wiley & Sons Ltd., Chichester, West Sussex, UK.

Yam, T.W. and Arditti, J. (2018) Orchid micropropagation: an overview of approaches and methodologies. In: Lee, Y.-I. and Yeung, E.C.-T. (eds.) *Orchid Propagation: From Laboratories to Greenhouses—Methods and Protocols*. Springer, New York, New York/USA, pp. 151–178.

Yam, T.W., Yeung, E.C., Ye, X.-L., Zee, S.-Y. and Arditti, J. (2002) Embryology-Seeds. In: Kull, T. and Arditti, J. (eds.) *Orchid Biology: Reviews and Perspectives*, VIII. Springer, Dordrecht, Netherlands, pp. 287–385.

Zhang, G.Q., Liu, K.W., Li, Z., Lohaus, R., Hsiao, Y.Y. *et al.* (2017) The *Apostasia* genome and the evolution of orchids. *Nature* 549, 379–383.

7 Micropropagation and Genetic Transformation of Ornamental Succulent Plants

Ayu Oshin Yap Sinaga[1,2], Yung Lun Lin[1], Wen-Hui Wu[1] and I-Chun Pan[1,3,4]*
[1]*Department of Horticulture, National Chung Hsing University, Taichung City, Taiwan;* [2]*Department of Biology, Institut Teknologi Sumatera, Lampung, Indonesia;* [3]*Advanced Plant and Food Crop Biotechnology Center, National Chung Hsing University, Taiwan;* [4]*Organic Agriculture Promotion Center, National Chung Hsing University, Taiwan*

Abstract

Ornamental succulents store water in their tissues, allowing them to thrive in drought conditions. They are valuable economically and environmentally, fitting well with eco-friendly designs like indoor spaces. However, their slow growth makes them vulnerable to overcollection and trafficking. Tissue culture can aid in their propagation while genetic transformation may enhance traits such as drought and salinity resistance and boost the synthesis of valuable compounds. By highlighting successful micropropagation examples, we aim to motivate universities and companies to invest in these plants.

Keywords: Micropropagation, ornamental horticulture, succulent plants, tissue culture media, plant growth regulators, transfection.

7.1 Introduction

Ornamental succulent plants are highly valued for their unique aesthetic appeal, low maintenance requirements, and adaptability to various environments, making them popular choices for landscaping, interior decoration, and urban gardening (Francini *et al.*, 2022; Orlóci and Fekete, 2023). With the global trend toward sustainable and minimalist living, the demand for succulents has surged, driven by their ability to thrive in water-scarce conditions and compatibility with eco-friendly design concepts. Succulence, defined as the ability of living cells to store substantial amounts of water for later use, enables these plants to maintain metabolic activity independent of external water sources. To be classified as succulent, a plant must store water in living tissues, make the stored water accessible for use, and sustain activity during periods of water scarcity (Eggli and Nyffeler, 2009). While succulents make up only about 3–5% of flowering plants, this represents a significant diversity, spanning up to 83 plant families and 12,500 species (Griffiths and Males, 2017; Nyffeler and Eggli, 2010). The evolution of succulence

*Corresponding author: icp@dragon.nchu.edu.tw

 Biotechnology of Orchids and Other Ornamentals (Eds P. Bolaños-Villegas and F-C. Chen)
DOI: 10.1079/9781800627895.0007

appears to have occurred across multiple lineages without a strong evolutionary pattern, demonstrating its adaptability. Succulents are characterized by their thick, fleshy leaves and stems adapted for water storage, with notable representation in families such as Aizoaceae, Apocynaceae, Asparagaceae, Asphodelaceae, Cactaceae, and Crassulaceae. The diverse adaptations of succulent plants, particularly their ability to store water, not only contribute to their popularity as ornamental plants but also underline their ecological importance. Understanding the mechanisms behind succulence provides valuable insights into how these plants thrive in arid conditions, emphasizing the importance of water storage for their survival and adaptability.

The ability of succulent plants to store water helps slow tissue water deficits during the day, buffer against daily fluctuations in water loss, and adapt to seasonal variations in water scarcity. This capacity effectively mitigates water-deficit stress, emphasizing the critical function of succulence in plant survival (Eggli and Nyffeler, 2009). Some desert succulents can reversibly lose up to 50% of their stored water during prolonged dry periods (Nobel, 2003). In species utilizing crassulacean acid metabolism (CAM), more than half of the water lost through nocturnal transpiration, due to stomatal opening at night, is drawn from stored reserves and replenished during the day when stomata remain closed (Schulte and Nobel, 1989). Succulent plants are commonly associated with ecological conditions of seasonal water availability but are less prevalent in regions with prolonged drought (Eggli and Nyffeler, 2009; Miriti *et al.*, 2007; Von Willert *et al.*, 1990). Their geographical distribution is heavily influenced by regional precipitation patterns, which are essential for the long-term survival and proliferation of species relying on this adaptation. Succulence is not confined to any specific mode of photosynthesis, appearing in different carbon fixation pathways such as the C3, in which carbon dioxide is initially converted into a triple carbon compound called 3-phosphoglycerate (hence C3), C4, in which CO_2 is converted into a compound with four carbons called oxaloacetate (hence C4), and in CAM plants. CAM, distinct from C4 photosynthesis, involves the nocturnal storage of malic acid, which is converted to CO_2 for daytime photosynthesis (Kluge and Ting, 1978). Tissue succulence supports this nocturnal storage of malic acid (Winter and Smith, 2022). Although CAM and succulence are often considered interdependent traits, recent studies in *Clusia* species with varying leaf anatomies suggest that CAM and the hydraulic capacitance provided by succulence are independent adaptations to drought stress (Leverett *et al.*, 2023). These adaptive traits of succulence, including the ability to mitigate water-deficit stress and support diverse modes of photosynthesis, underscore the ecological and functional importance of succulent plants. To further enhance their resilience, productivity, and ornamental value, the development of advanced techniques has become important, enabling large-scale propagation and precise genetic modifications to meet both commercial and conservation demands.

Conventional propagation of ornamental succulent plants is primarily achieved through methods such as seed propagation, cuttings, and grafting, which have been widely used for their simplicity. Seed propagation is a common and straightforward method; however, it can be challenging to obtain seedlings for certain species due to limitations in flowering, reproduction, and seed formation (Lema-Rumińska and Kulus, 2014). Additionally, seed propagation does not ensure genetic stability or true-to-type progenies due to the prevalence of cross-pollination and genetic variability. In some species, germination rates are low, often hindered by physical dormancy caused by a hard seed coat. For example, in *Opuntia ficus-indica*, seed propagation resulted in a germination rate as low as 5.3% (Stambouli-Essassi *et al.*, 2017). Moreover, plants derived from these seeds required five years to develop into mature plants, with the first flowering event occurring only after seven years. While vegetative propagation methods such as cutting and grafting offer advantages over seed germination, they are primarily limited to columnar species (Bouzroud *et al.*, 2022). Additionally, the propagation rate can be a constraint. These pressures have significantly contributed to the extinction risks of local populations and drastic shifts in the distribution and abundance of cacti. These challenges are further compounded by their slow growth rates and reliance on mature, reproductive individuals for population sustainability. Moreover, the limited number of plants produced through conventional methods remains insufficient to meet the growing global market demand (Lema-Rumińska and Kulus, 2014). Therefore, advanced propagation techniques, such as micropropagation and genetic transformation, can be utilized to meet the growing demand for succulent plants in horticulture and conservation efforts.

Micropropagation and genetic transformation are important tools for advancing the cultivation and conservation of ornamental succulent plants. Since the 1960s, there has been a fluctuating but increasing trend in the literature on ornamental succulent tissue culture (Fig. 7.1A). The number of publications grew steadily from 1960 until the 2000s, with a notable surge peaking in 2017 at 17 publications. This rising interest underscores the significance of tissue culture techniques in propagating ornamental succulents and provides a robust foundation for understanding the mechanisms underlying their growth. The expanding body of research reflects the growing demand to address challenges related to ornamental succulent cultivation and conservation. Similarly, research on the genetic transformation of ornamental succulents has shown a steady increase since the 1960s, with fluctuating trends over the years (Fig. 7.1B). The number of publications grew consistently from 1960 until the 2000s, peaking in 2017 at 12 publications. This growing interest underscores the importance of gene transformation techniques in improving the quality and quantity of ornamental succulents, offering solutions to enhance their traits and resilience. Figure 7.1C illustrates well the interdisciplinary nature of research in ornamental succulent tissue culture, with significant contributions from agriculture, biochemistry, and pharmacology, emphasizing their importance in medicinal applications and conservation. Publications in agricultural and

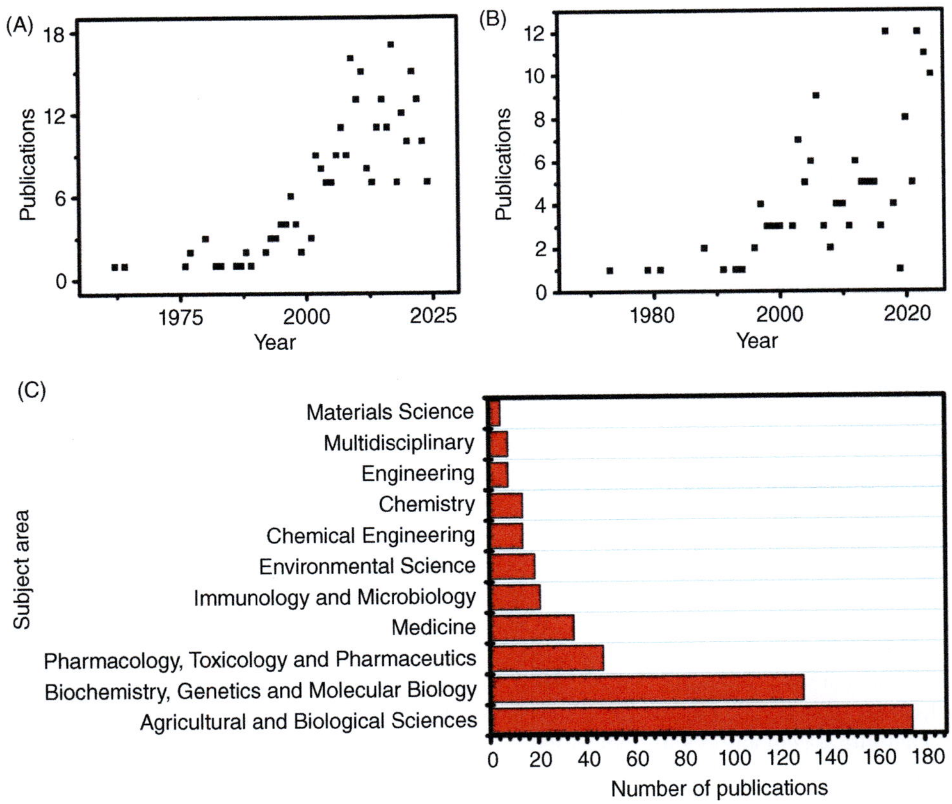

Fig. 7.1. Statistics related to publications in the field of propagation of succulent plants for the last 50 years. A) Number of publications related to succulent tissue culture. B) Number of publications related to gene transformation in succulents. C) Publications by subject (retrieved from the SCOPUS database).

biological sciences, as well as environmental science, focus on plant growth, adaptation mechanisms, and stress responses. Meanwhile, biochemistry research explores metabolic pathways, genetic expression, and molecular mechanisms underlying stress tolerance, CAM photosynthesis, and secondary metabolite production. Collectively, this research underscores the importance of tissue culture in advancing the propagation, conservation, and functional understanding of ornamental succulent plants.

The aim of this review is to compare and summarize conventional propagation methods with micropropagation and transformation techniques for plant improvement. Several ornamental succulent plant families, including Aizoaceae, Apocynaceae, Asparagaceae, Asphodelaceae, Cactaceae, and Crassulaceae, are examined to evaluate the differences in micropropagation and transformation methods. Key aspects such as explants, plant growth regulators, and the yields obtained through each micropropagation method are discussed. Additionally, the transformation methods, target genes, and promoters used are analyzed. The advantages and challenges associated with these techniques are also explored. A comprehensive understanding of micropropagation and transformation is a necessity for developing high-quality traits to meet the increasing market demand for ornamental succulent plants. This review provides a thorough overview of the current state of ornamental plant tissue culture, offering valuable insights to guide future research and development in the field.

7.2 Micropropagation of Ornamental Succulent Plants

The multiplication of ornamental succulent plants holds significant importance for both commercial and ecological purposes. As succulents continue to rise in popularity due to their aesthetic appeal, low maintenance, and ability to thrive in arid conditions, efficient propagation methods are indispensable to meet the growing market demand (Ong *et al.*, 2022). Multiplication of ornamental succulent plants allows for large-scale production of high-quality, uniform plants, ensuring consistent supply for ornamental, medicinal, and landscaping uses (Chavan *et al.*, 2013; Li *et al.*, 2023). Moreover, rapid multiplication is essential for conserving rare and endangered succulent species, helping preserve biodiversity and protect plants from overexploitation in their natural habitats (Bertsouklis *et al.*, 2024; Gupta *et al.*, 2024). By enabling sustainable production, multiplication techniques contribute to both economic growth in the horticulture industry and the long-term preservation of these unique and resilient plants.

Traditionally, ornamental succulent plants were primarily propagated vegetatively through methods such as cuttings and grafting, or by using seeds (Newland *et al.*, 1981; Roberto and Colombo, 2020). Vegetative propagation, including the use of stem or leaf cuttings, has been a popular approach due to its simplicity and ability to produce genetically identical plants with desirable traits. Grafting, another common technique, is particularly useful for propagating slow-growing or delicate species, as it enhances growth rates and improves resilience by attaching them to stronger rootstocks. Seed propagation, while less labor-intensive, is often slower and may result in genetic variability, making it less suitable for maintaining uniformity in ornamental succulents. Although these traditional methods are effective, they are time-consuming and limited in scale, driving the need for more advanced propagation techniques, such as tissue culture, to meet the increasing demand for succulent plants in horticulture and conservation efforts.

Micropropagation is a modern plant tissue culture technique used for the rapid and large-scale production of ornamental succulent plants (Mehbub *et al.*, 2022; Pandey and Pandey, 2018). This method involves explant preparation, culture initiation, multiplication, shooting and rooting, and acclimatization (Fig. 7.2). The comparison between conventional and micropropagation techniques reveals several key differences in terms of efficiency, cost, scalability, and labor intensity (Table 7.1). Conventional techniques, such as cuttings and grafting, typically require more time for plant growth and propagation, as they depend on slower, natural processes. In contrast, micropropagation allows for rapid plant multiplication, significantly reducing the time needed to produce large numbers of plants. While conventional methods may have lower upfront costs, they can become more expensive

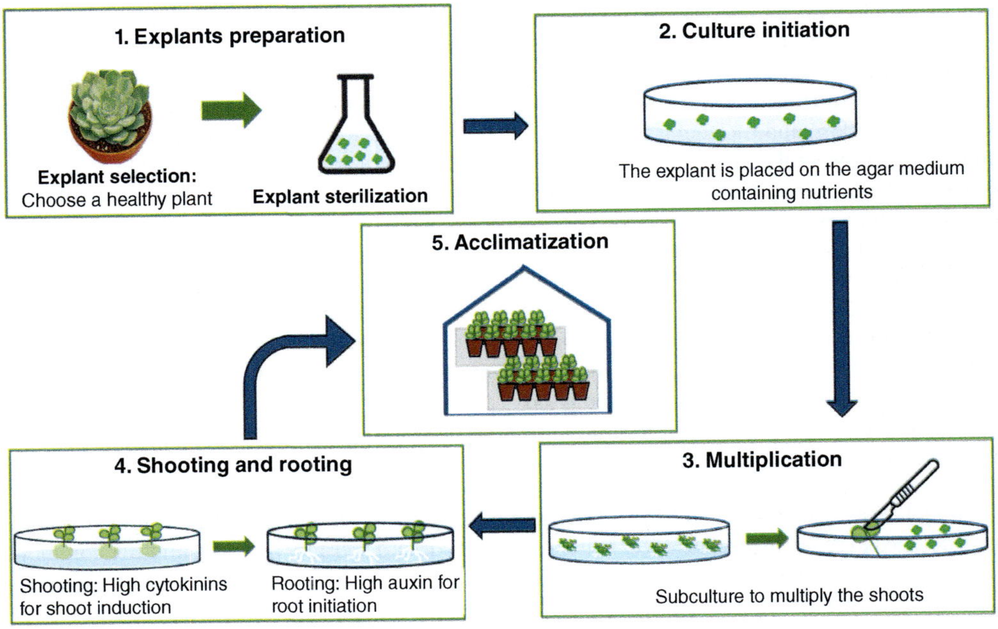

Fig. 7.2. General stages of micropropagation for ornamental succulent plants. Author's own image.

Table 7.1. Comparison of conventional and micropropagation techniques in succulent plants. Author's own table.

Factor	Conventional techniques	Micropropagation techniques
Time required	Slow; takes longer for plant growth and propagation.	Quick; allows rapid production and multiplication of plants.
Cost	Low upfront costs, but high in the long term due to inefficiency.	High initial setup costs but cost-effective eventually due to quick multiplication and higher yield.
Scalability	Limited scalability due to space, time, and resource constraints.	Highly scalable; suitable for large-scale production.
Risk of contamination	Substantial risk of pests and diseases, especially in field-based propagation.	Negligible risk of contamination, as the process occurs in sterile laboratory conditions.
Labor intensity	High labor intensity; requires manual care and attention.	Low labor intensity: much of the process is automated or requires little manual labor.

overall due to inefficiencies and the need for more manual labor. On the other hand, micropropagation involves higher initial setup costs but proves to be more cost-effective over time, especially for large-scale production. Scalability is another major advantage of micropropagation, as it allows for the efficient production of a large volume of plants, something that is challenging with conventional methods due to space, time, and resource limitations. Additionally, micropropagation carries a lower risk of contamination since the process takes place in controlled, sterile laboratory conditions, whereas conventional methods are more susceptible to pests and diseases in field settings. Finally, micropropagation reduces labor intensity, as much of the process is automated or requires less manual intervention, while conventional techniques demand more hands-on care and attention. Micropropagation offers significant advantages in terms of speed, scalability, and reduced labor, making it a more efficient method for large-scale and disease-free plant production.

7.2.1 Development of micropropagation techniques for ornamental succulents

The development of current ornamental succulent tissue culture protocols enables large-scale production, genetic improvement, and preservation of rare or endangered species. The success of succulent tissue culture relies heavily on critical factors such as explant selection, growth media composition, plant growth regulators, and environmental conditions like light and temperature. Explant selection plays a fundamental role in tissue culture, and tissues such as leaves, shoot tips, seeds, stem, or specialized structures such as phylloclades that must be chosen carefully for their high regenerative potential and adaptability to *in vitro* conditions. The preferred growth medium is Murashige and Skoog (MS), which provides all the necessary macro- and micronutrients to support cell division, elongation, and tissue growth, with the addition of plant growth regulators such as auxins and cytokinins to direct developmental pathways. Furthermore, the control of light intensity and temperature to mimic the plant's natural habitat is important to optimize growth. Species adapted to high-light or warm climates require different *in vitro* conditions compared to those thriving in cooler, low-light environments. By integrating these factors, researchers ensure efficient propagation, enhanced regeneration, and the long-term sustainability of ornamental succulent species.

In ornamental succulent tissue culture, the choice of explant is pivotal in determining the success of propagation and regeneration. Several types of explants have been utilized across different succulent families to initiate tissue culture Table 7.2. For instance, in *Aizoaceae*, explants such as leaves, callus, shoot tips, stems, and seeds are commonly used. Similarly, *Apocynaceae* employs leaves, stems, and shoot tips as primary explants. The *Asparagaceae* family utilizes diverse explants, including meristems, zygotic embryos, bulbils, stolons, shoots, leaves, and cotyledons, demonstrating its adaptability in tissue culture. For *Asphodelaceae*, offshoots, seeds, leaves, and inflorescences serve as suitable explants, while *Cactaceae* focuses on seeds, unopened flower buds, stem nodes, and mature phylloclades due to their robust regenerative capacity. The *Crassulaceae* family relies on leaves, stems, roots, flower buds, and apical nodes for tissue culture. In general, the most widely used explants in succulent tissue culture include seeds, shoot tips, leaves, bulbils, and stem nodes. These tissues are selected based on their capacity to dedifferentiate and regenerate into new plants under suitable *in vitro* conditions. Explants such as zygotic embryos and meristems are also favored for their high totipotency and reduced risk of contamination. Additionally, specialized structures like inflorescences and phylloclades are explored for their unique regenerative properties in specific species.

Aseptic sterilization of explants is a vital step in plant tissue culture to guarantee successful *in vitro* propagation. The process involves the use of sterilizing agents such as ethanol, sodium hypochlorite, or hydrogen peroxide to eliminate microbial contaminants from the surface of explants without damaging the tissue's viability (Misra and Misra, 2012). Proper sterilization not only prevents the growth of bacteria, fungi, or viruses in the culture medium but also maintains the aseptic environment required for optimal plant cell growth. The choice and concentration of sterilizing agents depend on the type of explant, its size, and its sensitivity to chemicals. Typically, the sterilization protocol includes an initial rinse with running tap water, followed by surface sterilization with a disinfectant solution, and a final rinsing step with sterile distilled water to remove any residual chemicals. This procedure ensures that the explant is free from contaminants, enabling healthy and uncontaminated growth during tissue culture processes.

Besides explant selection, the choice of growth medium is a critical factor that influences the success of propagation, regeneration, and growth in ornamental succulent tissue culture. Among the various media available, the MS medium is the most commonly used for succulent tissue culture due to its balanced composition of macro- and micro-nutrients, which support a wide range of plant species. The high nitrate, potassium, and ammonium content in MS medium is particularly beneficial for promoting

Table 7.2. Summary of micropropagation techniques for ornamental succulent plants. Author's own table.

Species	Explants used	Medium	Plant growth regulator	Light	Temperature	Reference
Aizoaceae						
Argyroderma testiculare	Leaves and callus	Callus induction: MS, 30 g l^{-1} sucrose, 7.5 g l^{-1} agar (pH 5.7) Shooting: ½ MS, 7.5 g l^{-1} sucrose, 1.8 g l^{-1} agar (pH 5.7)	A. Callus induction: 10 μM BA + 20 μM IAA B. Shooting: 10 μM BA C. Multiplication: 10 μM BA + 10 μM IAA	950 μmol m^{-2} s^{-1}	25 ± 2°C	Ofisi, 2017
Carpobrotus edulis	Shoot and leaves	Seedling: MS (vit), 3% sucrose, 7 g l^{-1} agar (pH 5.7) Shooting: MS (vit), 3% sucrose, 6 g l^{-1} agar (pH 5.7) Rooting: MS, 3% sucrose, 7 g l^{-1} agar (pH 5.7)	A. Shooting: 0.05 μM TDZ B. Rooting: 0.4 mg l^{-1} NAA	4000 lux	22 ± 2°C	Khattab and El Sherif, 2011
Drosanthemum hallii	Stem nodes and leaves	MS (vit), 3% sucrose, 7.5 g l^{-1} agar (pH 5.7)	A. Callus induction: 30 μM of 2-isopentyladenine (2iP) B. Shooting: 10 μM kinetin C. Rooting: PGR-free	4000 lux	25 ± 2°C	Mlungwana, 2018
Sceletium tortuosum	Seed	Seedling: MS salts, 10g l^{-1} agar (pH 5.8) Callus induction: MS (vit), 0.1 g l^{-1} myo-inositol, 30 g l^{-1} sucrose, 10 g l^{-1} (pH 5.8) Shoot induction: MS (vit), 0.1 g l^{-1} myo-inositol, 30 g l^{-1} sucrose, 10 g l^{-1} agar (pH 5.8)	A. Callus induction: 5 mg l^{-1} BA B. Shooting: 1 mg l^{-1} BA	38.27 μmol m^{-2} s^{-1}	22 ± 2°C	Makunga *et al.*, 2022
Sesuvium portulacastrum L.	Seed	Seedling: MS, 30 g l^{-1} sucrose, 8 g l^{-1} agar (pH 5.8) Rooting: ½ MS, 30 g l^{-1} sucrose, 8 g l^{-1} agar (pH 5.8)	A. Shooting: 1.0 mg l^{-1} zeatin + 0.3 mg l^{-1} NAA B. Rooting: 0.6 mg l^{-1} IBA	25 μmol m^{-2} s^{-1}	25°C	He *et al.*, 2021
Apocynaceae						
Adenium obesum	Leaves and stem	MS, 30 g l^{-1} sucrose, 8 g l^{-1} agar (pH 5.8)	Callus induction: 1.5 mg l^{-1} NAA + 1 mg l^{-1} BAP	40–60 μmol m^{-2} s^{-1}	25 ± 1°C	Rasad *et al.*, 2015

Adenium obesum (Forssk.) Roem. and Schult.	Shoot	Seedling: MS basal medium Shooting: MS, 30 g l^{-1} sucrose, 8.2 g l^{-1} agar (pH 5.7)	A. Shooting: 22.2 μM BA B. Rooting: 0.3% activated charcoal	20 μmol m^{-2} s^{-1}	25 ± 1°C	Kanchanapoom *et al.*, 2010
Huernia hystrix	Stem	MS, 30 g l^{-1} sucrose, 0.1 g l^{-1} myo-inositol, 0.8% agar	A. Shooting: 5.37 μM NAA + 22.19 μM BA B. Rooting: 1 μM of IBA	30 μmol m^{-2} s^{-1}	25°C	Amoo *et al.*, 2009
Asparagaceae						
Agave americana	Meristem	MS, 30 g l^{-1} sucrose, 2.5 g l^{-1} phytagel (pH 5.7)	A. Callus induction: 0.11 μM 2,4-D + 58.7 μM BA B. Shooting: PGR-free C. Rooting: PGR-free	35 μmol m^{-2} s^{-1}	25°C	Lecona-Guzmán *et al.*, 2017
Agave americana	Meristem	MS salts, 3% (w/v) sucrose, L2 vitamins	A. Callus induction: 2 mg l^{-1} 2,4-D B. Multiplication: 3 mg l^{-1} BA + 0.1 mg l^{-1} NAA	45 μmol m^{-2} s^{-1}	25 ± 2°C	Nazir *et al.*, 2019
Agave angustifolia Haw.	Mature zygotic embryos	Callus induction: MS, L2 vitamin, 60g l^{-1} sucrose Somatic embryos: 1/4 MS, L2 vitamin, 60 g l^{-1} sucrose	A. Callus induction: 13.59 μM 2,4-D + 4.44 μM BA B. Somatic embryos: 13.59 μM 2,4-D and 4.44 μM BA	Dark	25 ± 2°C	Reyes-Díaz *et al.*, 2017
Agave arizonica Gentry and Weber	Bulbils	Callus induction: MS, 555 μM inositol, thiamine-HCl 1.2 μM, sucrose 58.5 mM Shooting: MS, vitamin, sucrose 87.7 mM Rooting: MS, 58.5 mM sucrose	A. Callus induction: 1.4 μM 2,4-D B. Shooting: 44.4 μM BA and 0.5 μM NAA C. Rooting: PGR-free	25 μmol m^{-2} s^{-1}	25–28°C	Powers and Backhaus, 1989
Agave eantala Roxb. *Agave fourcroydes* Lem. *Agave sisalana* Perrine	Stolons	Multiplication: MS basal, 2% (w/v) sucrose, 10% (v/v) coconut water, 0.8% (w/v) agar Rooting: MS	A. Shooting: 0.075 mg l^{-1} NAA+ 0.1 mg l^{-1} IBA+ 0.5 mg l^{-1} kinetin B. Rooting: PGR-free	2000 lux	28°C	Binh *et al.*, 1990
Agave guiengola Gentry	Shoot	Seedling: MS basal medium Multiplication: MS, 30 g l^{-1} sucrose, 9 g l^{-1} agar pH5.7	Multiplication: 2 mg l^{-1} BA and 2 mg l^{-1} (2iP)	55 μmol m^{-2} s^{-1} (16 h light)	25 ± 2°C	Chávez-Ortiz *et al.*, 2021
Agave hybrid No. 11648	Leaves	Callus induction: SH medium, 6.5 mg l^{-1} carrageenan (pH 5.8) Shooting: SH medium, 6.5 g l^{-1} carrageenan, 30 g l^{-1} sucrose Rooting: SH medium	A. Callus induction: 3.0 mg l^{-1} BA + 0.5 mg l^{-1} NAA B. Shooting: 1.5 mg l^{-1} BA + 0.5 mg l^{-1} NAA C. Rooting: 0.1 mg l^{-1} IAA	40 mmol m^{-2} s^{-1} (12 h light)	28°C	Gao *et al.*, 2014

Continued

Table 7.2. Continued.

Species	Explants used	Medium	Plant growth regulator	Light	Temperature	Reference
Agave tequilana	Meristem and leaves	MS, 9.4 mM KNO_3 and 10.3 mM NH_4NO_3	A. Callus induction: 44 µM BA + 0.11 µM 2,4-D B. Rooting: PGR-free	50 µmol m^{-2} s^{-1}	25 ± 2°C	Valenzuela-Sánchez *et al.*, 2006
Agave tequilana Weber (cultivar azul)	Embryogenic callus	MS basal salts, 3% sucrose, 6 g l^{-1} phytagel	Multiplication: 9 µM 2,4-D + 1.3 µM BA	25 µmol m^{-2} s^{-1} (light 16 h)	27 ± 2°C	Santacruz-Ruvalcaba and Portillo, 2009
Agave vera-cruz Mill.	Shoot, cotyledons and leaves	MS salts with L2 vitamins, sugar, 0.8% agar	A. Callus induction: 4.52 µM 2,4-D or 5.37 µM NAA B. Somatic embryos: 5.37 µM NAA + 0.91 µM Zeatin	25 µmol m^{-2} s^{-1} (light 16 h)	25 ± 2°C	Tejavathi *et al.*, 2007
Asphodelaceae						
Aloe adigratana Reynolds	Offshoots	Shooting: MS, 30 g sucrose, 8 g Agar (pH 5.8) Rooting: ½ MS, 30 g sucrose, 8 g agar	A. Callus induction: 0.2 mg/L BAP + 0.2 mg l^{-1} NAA B. Shooting: 1.0 mg l^{-1} of BAP + 0.5mg l^{-1} of NAA C. Rooting: 1 mg l^{-1} of IBA or 0.5 of NAA	2000–2500 lux (fluorescent tube light)	25 ± 0.5°C	Niguse *et al.*, 2020
Aloe polyphylla	Seed	MS, 30 g l^{-1} sucrose, 0.1 g l^{-1} myo-inositol, 1% agar	A. Seedling: 0.5 mg l^{-1} IBA + 1.0 mg l^{-1} zeatin B. Shooting: 5 µM of mT ([6-(3-hydroxybenzylamino) purine]) + 1.0 mg l^{-1} zeatin	45 µmol m^{-2} s^{-1}	24 ± 1°C	Bairu *et al.*, 2007
Aloe vera L.	Shoot tip	MS, 30 g l^{-1} sucrose, 0.1% rosemary essential oil, 0.6% agar (pH 5.8)	A. Shooting: 0.2 mg l^{-1} IBA + 3mg l^{-1} BA+ 5% AvG (Aloe vera gel) B. Rooting: 0.2 mg l^{-1} IBA + 3mg l^{-1} BA + 10% AvG (Aloe vera gel)	36 µmol m^{-2} s^{-1}	22 ± 2°C	Hamdeni *et al.*, 2022
Aloe vera (L.) Burm. f	Leaves	MS (with vitamins), 3% sucrose, 0.8% (w/v) agar	A. Callus induction: 2 mg l^{-1} BAP + 0.5 mg l^{-1} NAA B. Shooting: 0.5 mg l^{-1} NAA C. Rooting: 1.5 mg l^{-1} BAP + 0.5 mg l^{-1} NAA	45 µmol m^{-2} s^{-1} (16 h light)	25 ± 2°C	Gupta *et al.*, 2024

Species	Explant	Medium	Plant growth regulators	Light	Temperature	Reference
Gasteria croucheri Bak.	Leaves	MS, 1% agar	A. Callus induction: 5 mg l^{-1} NAA + 2 mg l^{-1} BA B. Rooting:1 mg l^{-1} NAA+ 1 mg l^{-1} BA	35 µmol m^{-2} s^{-1}	25 ± 2°C	Bayley and Van Staden, 1987
Haworthia 'Sansenjyu' (*H. obtusa* × *H. comptoniana*)	Inflorescence	MS, 30g l^{-1} sucrose, 0.1g l^{-1} myo-inositol (pH 5.7)	A. Callus induction: 1.0 ppm BA + 0.1 ppm NAA B. Shooting: 2 ppm BA + –0.4 ppm NAA C. Rooting:0.4 ppm NAA + 0.4 ppm IBA	Callus induction: 10 µmol m^{-2} s^{-1} Shooting: 45 µmol m^{-2} s^{-1}	26°C	Chen *et al.*, 2019
Haworthia retusa	Leaves	MS, 3% sucrose, 0.8% agar (pH 5.8)	A. Somatic embryos: 16 µM IBA + 4 µM TDZ B. Shooting: 2 µM GA_3	45 µmol m^{-2} s^{-1}	25 ± 2°C	Kim *et al.*, 2019
Haworthia turgida Haw.	Leaves	MS, 3% sucrose (pH 5.8)	A. Callus induction: 1.0 ppm BA + 0.1 ppm NAA B. Shooting: 1 ppm BA + 0.2 ppm 2,4-D C. Rooting: 0.05 ppm NAA	40 µmol m^{-2} s^{-1}	24 ± 1°C	Liu *et al.*, 2017
Cactaceae						
Ariocarpus kotschoubeyanus (Lem.) K. Schum	Seed	Seedling: MS, 2 mg l^{-1} glycine, 100 mg l^{-1} myo-inositol, 30g l^{-1} sucrose Rooting: ½ MS, activated charcoal (1 g l^{-1})	A. Shooting: 13.3 µM BA + 5.4 µM NAA B. Somatic embryos: 8.9 µM BA+ 0.5-5.4 µM NAA	50 µmol m^{-2} s^{-1} (16 h light)	26 ± 2°C	Moebius-Goldammer *et al.*, 2003
Astrophytum asterias (Zucc.) Lem.	Seed	MS, Pierik medium (P2), agar	Multiplication: 1.0–1.5 mg l^{-1} BA	2000 lux (16 h light)	22–26°C	Ivannikov *et al.*, 2022
Blossfeldia liliputana Werderm	Seed	MS, Pierik medium (P2), agar	Multiplication: 1.0 mg l^{-1} BA + 0.1 mg l^{-1} 2,4-D	2000 lux (16 h light)	22–26°C	Ivannikov *et al.*, 2022
Mammillaria albicoma Boedeker *M. carmenae* Castañeda and Nunez *M. schiedeana* Ehrenberg	Unopened flower buds	MS, 3 g l^{-1} sucrose, phytagel	A. Shooting 5 mg l^{-1} BA + 0.1 mg l^{-1} NAA B. Rooting: 0.01 mg l^{-1} NAA	55 µmol m^{-2} s^{-1}	20°C	Wyka *et al.*, 2009)
Micranthocereus flaviflorus subsp. Densiflorus *Micranthocereus polyanthus* subsp. *alvinii*	Stem node	½ MS, 1.5% sucrose, 6 g agar	Multiplication: 1.34 µmol l^{-1} NAA	60 µmol m^{-2} s^{-1}	25 ± 3°C	Civatti *et al.*, 2017

Continued

Table 7.2. Continued.

Species	Explants used	Medium	Plant growth regulator	Light	Temperature	Reference
Notocactus magnificus	Seed	MS, 3% sucrose, 0.6% Agar	Multiplication: 22.2 µM BA	47 µmol m^{-2} s^{-1}	25 ± 2°C	de Medeiros *et al.*, 2006
Schlumbergera truncata cv. Russian Dancer	Mature phylloclades	Callus induction: MS salts, Staba vitamins, 3% w/v sucrose, 3 g l^{-1} phytagel Somatic embryos: MS salts, 0.1 g l^{-1}, myo-inositol, 7.0 µM kinetin, 3% w/v sucrose.	A. Callus induction: 22.7 µM TDZ + 1.3 µM NAA B. Somatic embryos: 26.6 µM BA, 27.2 µM TDZ	66 µmol m^{-2} s^{-1} (17 h light)	Callus initiation: 25–28°C Somatic embryo induction: 27–29°C	Al-Ramamneh *et al.*, 2006
Crassulaceae						
Crassula portulacea	Leaves	MS basal media	Multiplication: 13.2 µM BAP + 0.54 µM NAA	35 µmol m^{-2} s^{-1}	25 ± 1°C	Zhang *et al.*, 2021
Hylotelephium sieboldii *Hylotelephium sieboldii* var. *ettyuense*	Leaves, stems, roots, and flower buds	MS, 30 g l^{-1} sucrose, 2 g l^{-1} Gellan gum (pH 5.6-5.8)	Multiplication: 1 mg l^{-1} of NAA + 1 mg l^{-1} BA	50 µmol m^{-2} s^{-1}	25°C	Nakano *et al.*, 2005
Kalanchoe peltata *K. laxiflora* *K. tubiflora* *K. marmorata*	Leaves	MS salts, MS (vit), 30 g l^{-1} sucrose, 100 mg l^{-1} myo-inositol, 6.5 g l^{-1} Bacto-agar (pH5.75)	Shooting: 0.5 mg l^{-1} TDZ + 0.04 mg l^{-1} IAA	17 h light, 7h dark	25°C	Frello *et al.*, 2002
Kalanchoe blossfeldiana	Apical node and leaves	MS	A. Calli induction: 2 mg l^{-1} 2,4-D + 2 mg l^{-1} BAP B. Shooting: 0.4 mg l^{-1} BAP C. Rooting: 1 mg l^{-1} IBA	2000–3000 lux (16 h light)	24 ± 2	Kale *et al.*, 2018

cell division and tissue growth (Gamborg *et al.*, 1976), making it versatile for various explants, including leaves, stems, and seeds. Additionally, the availability of well-established protocols for MS medium further facilitates its widespread use in tissue culture laboratories. The easy availability of commercial MS medium products is also a reason for its widespread use in plant tissue culture. In some cases, another media such as Schenk and Hildebrandt (SH) or Pierik medium (P2) are preferred for specific plant species. SH medium served as the base medium for callus induction, shoot differentiation, and rooting, with various hormonal supplements (Schenk and Hildebrandt, 1972). In the case of *Agave* hybrid No. 11648, it was used to optimize nutrient uptake for zygotic embryos and meristems (Gao *et al.*, 2014). Meanwhile, P2 medium, with its modified nutrient composition, is used for the cultivation of stem explants, particularly in studies on callogenesis and plant cell haemogenesis (Pierik, 1997). In succulent plants such as *Astrophytum asterias* (Zucc.) Lem. and *Blossfeldia liliputana* Werderm., the P2 medium has been used for the growth of seed explants (Ivannikov *et al.*, 2022).

Some methods also include the addition of myo-inositol into the medium as an additive. Myo-inositol is synthesized from glucose-1-phosphate through an oxidation/reduction reaction catalyzed by NAD (nicotinamide adenine dinucleotide). It is a cyclic alcohol containing six hydroxyl groups and is one of several stereoisomers. Myo-inositol plays an important role in various biological processes in plants, including the biosynthesis of the plant cell wall, salt tolerance, phosphate storage, cell-to-cell communication, and the storage and transport of plant hormones (Loewus and Murthy, 2000). Many succulent varieties including *Sceletium tortuosum*, *Huernia hystrix*, *Aloe polyphylla*, *Haworthia 'Sansenjyu'* (*H. obtusa* × *H. comptoniana*), *Ariocarpus kotschoubeyanus* (Lem.) K. Schum, *Schlumbergera truncata* cv. Russian Dancer, *Kalanchoe peltata*, *K. laxiflora*, *K. tubiflora*, and *K. marmorata* add myo-inositol to the medium to promote the regeneration of different tissues (Table 7.2).

In addition to the growth medium, plant growth hormones are indispensable in ornamental plant tissue culture, regulating key processes such as callus induction, shoot regeneration, and rooting. Based on Table 7.2, auxins such as indole-3-acetic acid (IAA), naphthaleneacetic acid (NAA), indole-3-butyric acid (IBA), and 2,4-dichlorophenoxyacetic acid (2,4-D) are widely used for promoting cell division and elongation, particularly during callus induction and rooting. Cytokinins, including benzyladenine (BA), thidiazuron (TDZ), kinetin, zeatin, mT (meta-topolin), and isopentenyl adenine (2iP) are primarily employed for shoot induction due to their ability to stimulate cell proliferation and differentiation. As shown in Table 7.2, BA and TDZ are the most frequently used cytokinins, as they are effective across multiple plant families such as Aizoaceae, Cactaceae, Crassulaceae, and Asparagaceae. In many protocols, the balance between auxin and cytokinin concentrations is key; high auxin levels favor rooting, while high cytokinin levels promote shoot regeneration. Among these hormones, BAP emerges as the most used growth regulator for shoot induction in ornamental plant tissue culture due to its versatility and effectiveness across diverse species.

The balance between auxin and cytokinin concentrations is fundamental in shaping the development of various ornamental succulent plants. When auxin levels are higher than cytokinin (auxin > cytokinin), it promotes rooting. This is because auxins such as IAA, NAA, and 2,4-D stimulate cell elongation and division, which is necessary for root development. On the other hand, when cytokinin concentrations exceed auxin levels (cytokinin > auxin), shoot regeneration is favored. Cytokinins such as BAP, TDZ, and kinetin promote cell division and differentiation, leading to the development of shoots and buds. When auxin and cytokinin are present in equal concentrations (auxin = cytokinin), it often results in balanced growth or callus formation without significant differentiation into roots or shoots. The use of 2,4-D results in the highest callus formation rate (5.6%), when compared to other auxin hormones such as NAA, IBA, picloram, and dicamba (Nazir *et al.*, 2019). The synthetic 2,4-D hormone is also used in tissue culture for several succulent plants, such as *Agave angustifolia* Haw., *Agave arizonica* Gentry and Weber, *Agave tequilana*, *Agave vera-cruz* Mill., and *Kalanchoe blossfeldiana*, for calli induction.

The concentration of auxins and cytokinins used in ornamental plant tissue culture varies depending on the specific developmental process, such as callus induction, shoot regeneration, or rooting. For succulent plant regeneration, auxins like IAA, NAA, and IBA are used within a range of 0.01–1 mg l^{-1} for rooting and callus induction. For instance, IBA at concentrations of 1 mg l^{-1} or NAA at 0.5 mg l^{-1} is frequently utilized for rooting in species *Aloe adigratana*. Similarly, IAA is often applied at 0.1 mg l^{-1} for rooting in *Agave* hybrid No. 11648. Auxins are also necessary in callus formation when combined with cytokinins, as seen in *Aloe vera* where cytokinin BAP (2 mg l^{-1}) and auxin NAA (0.5 mg l^{-1}) are both required, and in *Gasteria croucheri* for auxin NAA (5 mg l^{-1}) and for cytokinin BAP (2 mg l^{-1}). The ability of auxins to promote cell division and elongation is essential for these processes (Table 7.2).

Cytokinins, including BAP, TDZ, and kinetin, are primarily used to stimulate shoot regeneration. Table 7.2 shows a typical range of 0.4–5 mg l^{-1} for cytokinins, with BAP being the most widely used. For example, BAP at 0.4 mg l^{-1} was used in *Kalanchoe blossfeldiana*, while BAP at 5 mg l^{-1} combined with 0.1 mg l^{-1} NAA was used in *Mammillaria albicoma*, *M. carmenae*, and *M. schiedeana* to promote shoot induction. TDZ was used at lower concentrations, such as 0.05–0.5 mg l^{-1}, in species like *Opuntia ficus-indica* and *Crassula ovata* to enhance shoot proliferation. Kinetin was less commonly used, but it was effective at 0.5 mg l^{-1} for shoot induction in *Agave cantala* Roxb., *A. fourcroydes* Lem., and *A. sisalana* Perrine, while *Drosanthemum hallii* required 10 μM. The balance between auxins and cytokinins determines the outcome, with higher cytokinin concentrations favoring shoot regeneration, while higher auxin concentrations promote rooting and callus formation. This targeted use of hormones ensures successful propagation and regeneration in ornamental succulent plant tissue culture.

Apart from growth hormones, environmental conditions, including temperature and light intensity, are important for the successful tissue culture of ornamental succulents. Table 7.2 shows that the temperature range commonly used in succulent tissue culture varies between 20–28°C, depending on the family. For example, Aizoaceae and Asphodelaceae are cultured at 22–25°C and 22–26°C, respectively, while Cactaceae prefers a slightly cooler range of 20–27°C. Notably, *Mammillaria albicoma*, *M. carmenae*, and *M. schiedeana* achieved good regeneration at a relatively low temperature of 20°C. Light intensity also shows significant variation, with values ranging from as low as 10 μmol m^{-2} s^{-1} for shade-tolerant species like *Haworthia* to as high as 4000 lux for sun-adapted plants like *Drosanthemum hallii*. This variation in light and temperature conditions ensures that the culture environment supports optimal growth and regeneration, tailored to the needs of distinct species and families. Additionally, families like Asparagaceae and Crassulaceae, which are adapted to warm and moderate conditions, are cultured at 25–28°C and 24–25°C, respectively, with light intensities typically between 25 and 55 μmol m^{-2} s^{-1}. Meanwhile, light intensity might determine whether plant tissues undergo callus proliferation or shoot regeneration in succulent tissue culture. For *Haworthia* 'Sansenjyu', low light intensity (10 μmol m^{-2} s^{-1}) is more effective in inducing callus formation, while high light intensity (45 μmol m^{-2} s^{-1}) is more favorable for induction (Chen *et al.*, 2019).

An important consideration in succulent tissue culture is whether the environmental conditions *in vitro* match the original habitat of the plants. Many ornamental succulents, such as *Agave* species from Asparagaceae, thrive in hot, arid climates and are cultured at higher temperatures of 28°C with moderate light levels. Conversely, cooler-temperature succulents, such as *Mammillaria albicoma* from Cactaceae, are cultured at a lower temperature range of 20°C, reflecting their original habitat in arid regions that have extreme temperature differences between day and night. This adjustment is crucial, as providing conditions that mimic the plant's natural habitat can improve growth and regeneration *in vitro*. While some families may share similar temperature and light preferences, species-level adaptations often necessitate specific adjustments to ensure successful tissue culture outcomes. The careful calibration of temperature and light conditions allows researchers to cater to the diverse ecological adaptations of ornamental succulents, enhancing both propagation efficiency and plant health.

7.2.2 Challenges and prospects in micropropagation of ornamental succulent plants

Micropropagation of ornamental succulent plants faces several challenges that hinder its widespread application (Abdalla *et al.*, 2022). One significant issue is callus formation, which can delay organogenesis and result in somaclonal variation, reducing the genetic stability of the propagated plants. Another major challenge is contamination, especially during the establishment phase, as succulents' thick leaves and stems often harbor microorganisms that are difficult to eliminate. Additionally, succulents typically exhibit long processes under *in vitro* conditions, which prolongs the micropropagation cycle and reduces overall efficiency. These factors combined create a need for optimized protocols to ensure successful propagation and commercial scalability.

The prospects of ornamental succulent tissue culture may involve a revolution through the integration of nanotechnology, artificial intelligence (AI), and genetic engineering. Nanotechnology offers significant potential by enhancing nutrient delivery, improving plant growth, and reducing contamination through nanoparticle-based media supplements and antimicrobial agents (Gunasena *et al.*, 2024). For instance, nano-fertilizers and nano-silver particles can optimize nutrient uptake and protect cultures from microbial infections, addressing key challenges in tissue culture systems (Tung *et al.*, 2021a,b). Meanwhile, AI-driven technologies, such as machine learning algorithms, can analyze vast datasets to optimize culture conditions, predict growth patterns, and automate decision-making processes in real-time, enhancing efficiency and scalability (Hesami and Jones, 2020). Additionally, genetic engineering provides opportunities to introduce desirable traits such as enhanced growth rates, improved stress resistance, and disease tolerance, ensuring the production of high-quality, resilient succulents (Bapat *et al.*, 2023; Wijerathna-Yapa and Hiti-Bandaralage, 2023). The combination of these innovative technologies will not only streamline the micropropagation process but also enable the large-scale, sustainable production of ornamental succulents tailored to meet the demands of an evolving market.

7.3 Genetic Transformation in Ornamental Succulent Plants

7.3.1 Recent development of gene transformation

The transformation of ornamental succulents is driven by both practical and research-oriented goals, enhancing their adaptability, productivity, and commercial value. The purpose of gene transformation in ornamental succulent plants extends beyond aesthetic improvements to encompass functional enhancements and scientific advancements. Several transformation methods for ornamental succulents are summarized in Table 7.3. One primary goal is to improve stress tolerance, particularly drought and salinity resistance, as seen in the transformation of *Mesembryanthemum crystallinum* using genes whose expression is driven by the Cauliflower Mosaic Virus (*CaMV*) *35S* promoter and others (Hwang *et al.*, 2019). Another important objective is to enhance the plants' metabolic properties, such as boosting the production of bioactive compounds in medicinal succulents such as *Rhodiola rosea*, where genes involved in secondary metabolite biosynthesis are introduced (Himmelboe *et al.*, 2013). Transformation is also utilized for growth optimization, such as employing *Agrobacterium rhizogenes* to transfect and introduce *rol* genes in species such as *Kalanchoe blossfeldiana* (Christensen *et al.*, 2008; Thirukkumarana *et al.*, 2010) and *Agave* hybrids, improving root architecture and growth rates (Gao *et al.*, 2014). To achieve these goals, the transformation of ornamental succulent plants is fundamental, making the development of tissue culture-based transformation techniques particularly significant.

Gene transformation in ornamental succulent plants relies on a variety of explants to facilitate successful genetic modification. Commonly used explants include callus tissues, shoots, leaves, meristems, and roots, depending on the species and the transformation method employed. *Opuntia ficus-indica* has been successfully transformed using meristem explants through particle bombardment, achieving a transient expression efficiency of 34–36%. This method allows for the direct delivery of genetic material into plant cells, overcoming the limitations of traditional transformation techniques (Llamoca-Zárate *et al.*, 1999). Transformation of *Mesembryanthemum crystallinum* using *Agrobacterium tumefaciens* strain EHA105 resulted in a cotyledonary node transformation efficiency of 4.6% (Agarie *et al.*, 2020), while transformation using callus achieved 1.8–3.0% efficiency (Hwang *et al.*, 2019). In contrast, transformation via seedling roots reached up to 85% (Hwang

Table 7.3. Summary of gene transformation methods for ornamental succulent plants

Species	Explants	Transplantation method	Transgenic genes	Promoter	Purpose	Reference
Aizoaceae						
Mesembryanthemum crystallinum	Callus, root	*A. rhizogenes* and *A. tumefaciens*	*GUS*	CaMV 35S	Efficient transformation protocol	(Hwang *et al.*, 2019)
Mesembryanthemum crystallinum L.	Cotyledonary node	*Agrobacterium tumefaciens*	*Enhanced Green Fluorescent Protein (EGFP) and Hygromycin Phosphotransferase Gene (HPH)*	CaMV 35S	Efficient transformation protocol	(Agarie *et al.*, 2020)
Mesembryanthemum crystallinum L.	Seedling roots	*Agrobacterium rhizogenes*	*GFP*	CaMV 35S	Non-destructive detection in transgenic plant	(Konieczny *et al.*, 2011)
Asparagaceae						
Agave desmettiana	Meristems from bulbils	*Agrobacterium*	*GUS*	CaMV35S	Efficient transformation protocol	(Gutiérrez-Mora *et al.*, 2014)
Agave hybrid No. 11648 (H.11648) [(*A. amaniensis* 9 *A. angustifolia* Haw.) 9 *A. amaniensis*]	Callus and shoot	*Agrobacterium*	*GUS*	CaMV35S	Efficient transformation protocol	(Gao *et al.*, 2014)
Agave salmiana	Callus	*Agrobacterium* / Gene gun	*GUS*	CaMV 35 S	Reduce breeding time, increase yield and disease resistance	(Flores-Benítez *et al.*, 2007)
Sansevieria trifasciata	Leaves	*Agrobacterium tumefaciens* (Cut-dip-budding)	*Phytoene desaturase (PDS)*	CaMV 35S	CRISPR/Cas9 gene editing	(Lu *et al.*, 2024)

Asphodelaceae						
Aloe barbadensis Miller	Stem, sheaths, leaves	*Agrobacterium*	*GUS*	CaMV 35 S	Efficient transformation protocol	(He *et al.*, 2007)
Furcraea foetida	Freshly grown shoot tips	*Agrobacterium*	*GUS*	CaMV 35 S	Reduce breeding time, increase yield and disease resistance	(Naz *et al.*, 2022)
Cactaceae						
Notocactus scopa cv. Soonjung	whole plant	*Agrobacterium* (Vacuum penetration, injection, puncture)	*GUS*	CaMV 35S	Efficient transformation protocol	(Seol *et al.*, 2008)
Opuntia ficus-indica	Cladodes	*Agrobacterium tumefaciens*	*GUS*	CaMV 35S	Efficient transformation protocol	(Silos-Espino *et al.*, 2006)
Opuntia ficus-indica	Immature pistil/ filament	*Agrobacterium tumefaciens*	*TdTomato fluorescence*	CMP	Non-destructive detection in transgenic plant	(Felker *et al.*, 2018)
Opuntia ficus-indica	Meristems	Particle bombardment	*GUS*	CaMV35S	Efficient transformation protocol	(Llamoca-Zárate *et al.*, 1999)
Rhipsalidopsis cv. CB5	Callus	*Agrobacterium*	*GUS*	CaMV 35S	Control ethylene to extend flowering period	(Al-Ramamneh *et al.*, 2006)
Crassulaceae						
Crassula arborescens	Leaves	*Agrobacterium tumefaciens* (Cut-dip-budding)	*Phytoene desaturase* (*PDS*)	CaMV 35S	CRISPR/Cas9 gene editing	(Lu *et al.*, 2024)
Graptopetalum paraguayense	Hairy root	*Agrobacterium rhizogenes* (dip, vacuum, sonication, and puncture)	*GFP* and *RUBY*	CaMV35S	Genetic improvement and gene function research	(Zhou *et al.*, 2024)

Continued

Table 7.3. Continued.

Species	Explants	Transplantation method	Transgenic genes	Promoter	Purpose	Reference
Hylotelephium spectabile	Hairy root	*Agrobacterium rhizogenes* (dip, vacuum, sonication, and puncture)	*GFP* and *RUBY*	CaMV35S	Genetic improvement and gene function research	(Zhou *et al.*, 2024)
Kalanchoe blossfeldiana	Leaves	*Agrobacterium*	1. *gus* gene 2. *rol* genes (*rol A, B, C* and *D* genes) of Ri T-DNA 3. recombinase gene (R) 4. *lacZ* gene	CaMV 35S	Genetic improvement and gene function research	(Thirukkumarana *et al.*, 2010)
Kalanchoe blossfeldiana	Leaves	*Agrobacterium*	*rol* genes	N/A	Dwarf genotypes development	(Christensen *et al.*, 2008)
Kalanchoe blossfeldiana	Leaves	*Agrobacterium tumefaciens*	*Phytoene desaturase* (*PDS*)	CaMV 35S	CRISPR/Cas9 gene editing	(Lu *et al.*, 2024)
Kalanchoë blossfeldiana	Leaves	*Agrobacterium tumefaciens*	*rolB, rolC, ΔORF13a,* and *ORF14*	CaMV35S	Genetic improvement and gene function research	(Favero *et al.*, 2021)
Kalanchoe laxiflora	Leaves	*Agrobacterium*	*gusA, gusPlus*	CaMV35S	Efficient transformation protocol	(Wang *et al.*, 2019)
Kalanchoe pinnata 'Cathe- dral Bells'	Leaves	*Agrobacterium*	*ZsGreen1*	FMV35S	Non-destructive detection in transgenic plant	(Cho *et al.*, 2021)
Kalanchoe tetraphylla	Hairy root	*Agrobacterium rhizogenes* (dip, vacuum, sonication, and puncture)	*GFP* and *RUBY*	CaMV35S	Genetic improvement and gene function research	(Zhou *et al.*, 2024)
Cultivars of *Kalanchoe*: '57', 'Jackie', 'Jodie', 'Molly', 'Naomi', 'Sarah', 'Simone', 'Suzanne' and 'Yellow African'	Nodal cutting and leaves	*Agrobacterium tumefaciens*	*Arabidopsis SHI* (Short Internodes) gene (*AtSHI*)	CaMV35S	Dwarf genotypes development	(Lütken *et al.*, 2010)

Rhodiola rosea	Hairy root	*Rhizobium rhizogenes*	*rol* genes	T-DNA (pRiA4)	*Rhizobium rhizogenes*-mediated transformation	(Martínez *et al*., 2020)
Rhodiola rosea L.	Callus	*Agrobacterium*	*uidA(GUS)* and *GFP*	CaMV 35S	Efficient transformation protocol	(Mirmazloum *et al*., 2014)
Rhodiola pachyclados R. rosea	Stem	*Rhizobium rhizogenes*	*rol* genes	T-DNA (pRiA4)	Efficient transformation protocol	(Himmelboe *et al*., 2013)
Sedum alfredii	Hairy root	*Agrobacterium rhizogenes* (dip, vacuum, sonication, and puncture)	*GFP* and *RUBY*	CaMV35S	Efficient transformation protocol	(Zhou *et al*., 2024)

et al., 2019). Konieczny et al. (2011) utilized *Agrobacterium rhizogenes* strain ARqua1 to infect 7- and 14-day-old ice plant seedlings, achieving transformation efficiencies of 6% and 20%, respectively. The transformed roots exhibited a hairy root phenotype and were maintained in liquid MS medium without exogenous plant hormones for over two years. Additionally, transgenic callus was induced from these hairy roots and cultured on solidified MS medium supplemented with cytokinin and auxin for at least 4–6 weeks. Similarly, Hwang et al. (2019) successfully used *A. rhizogenes* to transiently transform seedling roots, obtaining 33–100% transient transformation efficiency in 3-day-old seedlings. Furthermore, pot-grown juvenile plants infected with *A. rhizogenes* were hydroponically grown to develop transformed roots expressing *GUS* reporter genes, with over 27% of plants infected with *A. rhizogenes* strain A8196 maintaining transformed roots for eight consecutive weeks (Hwang et al. 2019).

Agrobacterium transformation infection methods significantly impact plant transformation efficiency (Zhou et al., 2024). Among the tested methods, the dip method demonstrated the highest transformation efficiency, achieving a 50% positive rate, significantly outperforming other techniques such as vacuum infiltration, sonication, and tissue puncturing, which each had an efficiency of approximately 20% (Zhou et al., 2024). The low survival rate of leaves in these methods is attributed to cell death caused during the infection process, which hinders the plants' ability to recover and successfully integrate the introduced genetic material.

Apart from transformation methods, the use of transgenic genes in the transformation of ornamental succulent plants has opened up new possibilities for enhancing their ornamental and functional value. Commonly utilized transgenic genes include reporter genes such as *GUS* (β-glucuronidase) and *GFP* (green fluorescent protein), which are used to monitor and confirm successful transformation events. Additionally, *rol* genes derived from *Agrobacterium rhizogenes* (e.g., *rolA*, *rolB*, and *rolC*) are frequently introduced to improve root architecture and enhance growth rates which are particularly valuable for ornamental succulents. Fluorescent protein-related genes, such as *ZsGreen1*, have been successfully expressed in *Kalanchoe pinnata*, producing strong green fluorescence throughout the plant organs under the regulation of the Figwort Mosaic Virus (*FMV*) *35S* promoter (Cho et al., 2021). Additionally, transgenic *Kalanchoe* plants have been developed with *Short Internodes* (*SHI*) genes to induce dwarf phenotypes using *Agrobacterium*-mediated transformation with the *CaMV 35S* promoter. Overexpression of *AtSHI* in *Kalanchoe* resulted in a range of growth effects, from extremely dwarf plants to those lacking elongated inflorescences (Lütken et al., 2010). This genetic modification holds significant potential for horticultural applications in controlling plant growth.

Promoters are also required for gene transformation in ornamental succulent plants, as they regulate the expression of transgenes introduced into the plant's genome. The most used promoter is the *Cauliflower Mosaic Virus 35S* promoter (*CaMV 35S*), which is widely recognized for its strong and constitutive expression across various plant species (Table 7.3). This promoter has been effectively employed in multiple succulents, including *Mesembryanthemum crystallinum* (Aizoaceae) (Agarie et al., 2020; Hwang et al., 2019; Konieczny et al., 2011), *Agave salmiana*, *Agave desmettiana*, (Asphodelaceae) (Flores-Benítez et al., 2007; Gutiérrez-Mora et al., 2014), and *Kalanchoe blossfeldiana* (Crassulaceae) (Thirukkumarana et al., 2010), ensuring high levels of expression for transgenes like *GUS* and *GFP*. In addition to the *CaMV 35S* promoter, specialized promoters, such as *FMV35S*, and *ZsGreen1* (derived from *Zoanthus* sp.), have been employed to enhance the non-destructive detection of transformation plants (Cho et al., 2021). The choice of promoter significantly impacts the efficiency, stability, and spatial expression of transgenes, making it a critical factor in the success of genetic transformation in ornamental succulents.

7.3.2 Advances and challenges in genetic transformation

An efficient delivery system for gene editing tools is often a key limiting factor in the application of gene editing technologies (Lu et al., 2024). *Agrobacterium tumefaciens*-mediated transformation and particle bombardment are the primary methods used for plant genetic transformation. However, these approaches are not suitable for all plant species. In addition to the time-consuming and labor-intensive nature of tissue culture, the transformation efficiency of these methods varies significantly depending on the plant species and genotype, further restricting their applicability. Contamination during tissue culture, along with somaclonal variation, further reduces the success rate of producing genetically stable plants (Abdalla *et al.*, 2022). To address these issues, ongoing research is focusing on optimizing transformation protocols, using alternative explants like embryogenic callus, and incorporating technologies such as CRISPR/Cas9 for targeted and efficient gene delivery (Demirer *et al.*, 2021; Mujtaba *et al.*, 2021). The CRISPR/Cas system has emerged as a powerful tool for precise and efficient genome editing in plants, offering several advantages over traditional genetic modification techniques (Kumar and Jain, 2015). One of its major strengths is its simplicity and efficiency, as the system requires only a single-guide RNA (sgRNA) and the Cas9 protein to induce targeted double-strand breaks in DNA, which are then repaired by the cell's natural mechanisms. Additionally, its multiplexing capability allows researchers to edit multiple genes simultaneously, facilitating the study of complex genetic pathways and accelerating crop improvement programs. This feature is particularly beneficial for plants with polyploid genomes, where multiple copies of a gene need to be modified at the same time. Furthermore, CRISPR-based base editing and prime editing offer precise modifications without inducing double-strand breaks, reducing the risk of unintended mutations. By combining the CRISPR/Cas system with the cut-dip-budding (CDB) gene delivery system, researchers have developed a method for genetic transformation and gene editing of succulents without tissue culture, effectively overcoming low transformation efficiency (Lu *et al.*, 2024). These advancements highlight the ongoing efforts to improve genetic transformation efficiency and overcome species-specific limitations. Continued research and innovation will be pivotal in enhancing transformation success rates and expanding the applicability of gene editing technologies across a broader range of ornamental succulent plant species.

7.4 Conclusions

The micropropagation and genetic transformation of ornamental succulent plants present both significant opportunities and challenges. The development of ornamental succulent micropropagation techniques for families such as Aizoaceae, Apocynaceae, Asparagaceae, Asphodelaceae, Cactaceae, and Crassulaceae has made it easier for researchers and farmers to meet the growing demand. These plants, with their unique adaptations to arid environments, are in high demand for their aesthetic and functional value. Traditional propagation methods, while effective, are slow and limited in scalability, driving the development of advanced techniques such as tissue culture and genetic engineering. Despite the promise of these technologies, challenges like low transformation efficiency, contamination, and the difficulty of regenerating transformed plants due to succulent-specific traits remain obstacles. However, ongoing research is addressing these issues through innovations in growth regulators, nanotechnology, and artificial intelligence, which hold the potential to streamline propagation and improve transformation success rates.

Succulent plants not only possess distinctive morphological characteristics with high commercial and ornamental value but also exhibit exceptional heat tolerance, drought resistance, and high water-storage capacity. Additionally, they utilize diverse photosynthetic pathways, including C3, C4, and CAM. With the advancement of genome sequencing in succulents, the development of an efficient and stable regeneration and genetic transformation system is crucial. Such systems will facilitate the identification of key genes responsible for abiotic stress tolerance and water-storage capacity, as well as elucidate the physiological mechanisms underlying different photosynthetic modes. Leveraging genetic transformation and genome editing technologies can enhance ornamental traits, improve stress resilience, accelerate growth, and enable the production of valuable bioactive compounds. These advancements have broad applications in horticulture, pharmaceuticals, and sustainable agriculture. The future prospects for the propagation and genetic modification of ornamental succulents are promising, offering potential benefits not only for succulent breeding but also for the improvement of other crop species.

References

Abdalla, N., El-Ramady, H., Seliem, M.K., El-Mahrouk, M.E., Taha, N. *et al.* (2022) An academic and technical overview on plant micropropagation challenges. *Horticulturae* 8, 677.

Agarie, S., Umemoto, M., Sunagawa, H., Anai, T. and Cushman, J. C. (2020) An *Agrobacterium*-mediated transformation via organogenesis regeneration of a facultative CAM plant, the common ice plant *Mesembryanthemum crystallinum* L. *Plant Production Science* 23, 343–349.

Al-Ramamneh, E. A.-D., Sriskandarajah, S. and Serek, M. (2006) Plant regeneration via somatic embryogenesis in *Schlumbergera truncata*. *Plant Cell, Tissue and Organ Culture* 84, 333–342.

Amoo, S., Finnie, J. and Van Staden, J. (2009) *In vitro* propagation of *Huernia hystrix*: an endangered medicinal and ornamental succulent. *Plant Cell, Tissue and Organ Culture* 96, 273–278.

Bairu, M.W., Stirk, W.A., Dolezal, K. and Van Staden, J. (2007) Optimizing the micropropagation protocol for the endangered *Aloe polyphylla*: can meta-topolin and its derivatives serve as replacement for benzyladenine and zeatin? *Plant Cell, Tissue and Organ Culture* 90, 15–23.

Bapat, V.A., Kavi Kishor, P., Jalaja, N., Jain, S.M. and Penna, S. (2023) Plant cell cultures: biofactories for the production of bioactive compounds. *Agronomy* 13, 858.

Bayley, A.D. and Van Staden, J. (1987) Propagation of *Gasteria croucheri* Bak. from shoot producing callus. *Plant Cell, Tissue and Organ Culture* 11, 227–231.

Bertsouklis, K., Kartsonas, E. and Carra, A. (2024) Seed germination and micropropagation of ornamental plants. *Horticulturae* 10, 541.

Binh, L., Muoi, L., Oanh, H., Thang, T. and Phong, D. (1990) Rapid propagation of agave by in vitro tissue culture. *Plant Cell, Tissue and Organ Culture* 23, 67–70.

Bouzroud, S., El Maaiden, E., Sobeh, M., Devkota, K. P., Boukcim, H., Kouisni, L. and El Kharrassi, Y. (2022) Micropropagation of *Opuntia* and other cacti species through axillary shoot proliferation: a comprehensive review. *Frontiers in Plant Science* 13, 926653.

Chavan, J.J., Gaikwad, N.B. and Yadav, S.R. (2013) High multiplication frequency and genetic stability analysis of *Ceropegia panchganiensis*, a threatened ornamental plant of Western Ghats: conservation implications. *Scientia Horticulturae* 161, 134–142.

Chávez-Ortiz, L. I., Morales-Domínguez, J. F., Rodríguez-Sahagún, A. and Pérez-Molphe-Balch, E. (2021) *In vitro* propagation of *Agave guiengola* Gentry using semisolid medium and temporary immersion bioreactors. *Phyton* 90, 1003.

Chen, Y.-M., Huang, J.-Z., Hou, T.-W. and Pan, I.-C. (2019) Effects of light intensity and plant growth regulators on callus proliferation and shoot regeneration in the ornamental succulent *Haworthia*. *Botanical Studies* 60, 1–8.

Cho, K.H., Vieira, A.E., Kim, J.Y., Clark, D.G. and Colquhoun, T.A. (2021) Transformation of Kalanchoe pinnata by *Agrobacterium tumefaciens* with *ZsGreen1*. *Plant Cell, Tissue and Organ Culture* 146, 401–407.

Christensen, B., Sriskandarajah, S., Serek, M. and Müller, R. (2008). Transformation of *Kalanchoe blossfeldiana* with *rol*-genes is useful in molecular breeding towards compact growth. *Plant Cell Reproduction* 27, 1485–1495.

Civatti, L. M., Nazaré, M. and Bellintani, M. C. (2017). Micropropagation of two species of *Micranthocereus* (Cactaceae) with ornamental potential native to Bahia, Brazil. *African Journal of Biotechnology* 16, 749–762.

de Medeiros, L.A., de Ribeiro, R.C.S., Gallo, L.A., de Oliveira, E.T. and Dematté, M.E.S.P. (2006) *In vitro* propagation of *Notocactus magnificus*. *Plant Cell, Tissue and Organ Culture* 84, 165–169.

Demirer, G. S., Silva, T. N., Jackson, C. T., Thomas, J. B., W. Ehrhardt, D. *et al.* (2021) Nanotechnology to advance CRISPR–Cas genetic engineering of plants. *Nature Nanotechnology* 16, 243–250.

Eggli, U. and Nyffeler, R. (2009) Living under temporarily arid conditions-succulence as an adaptive strategy. *Bradleya* 27, 13–36.

Favero, B.T., Tan, Y., Lin, Y., Hansen, H.B., Shadmani, N. *et al.* (2021) Transgenic *Kalanchoë blossfeldiana*, containing individual rol genes and open reading frames under 35S promoter, exhibit compact habit, reduced plant growth, and altered ethylene tolerance in flowers. *Frontiers in Plant Science* 12, 672023.

Felker, P., Bunch, R., Tine, J.A., Russo, G.R., Gould, J. *et al.* (2018) Stable transformation of *Opuntia ficus* indica callus cultures as evidenced by fluorescence of the tandem dimer Tomato gene. *Journal of the Professional Association for Cactus Development* 20, 34.

Flores-Benítez, S., Jiménez-Bremont, J.F., Rosales-Mendoza, S., Argüello-Astorga, G. R., Castillo-Collazo, R., and Alpuche-Solís, Á.G. (2007) Genetic transformation of *Agave salmiana* by *Agrobacterium tumefaciens* and particle bombardment. *Plant Cell, Tissue and Organ Culture* 91, 215–224.

Francini, A., Romano, D., Toscano, S. and Ferrante, A. (2022) The contribution of ornamental plants to urban ecosystem services. *Earth* 3, 1258–1274.

Frello, S., Venerus, E. and Serek, M. (2002) Regeneration of various species of *Crassulaceae*, with special reference to Kalanchoe. *The Journal of Horticultural Science and Biotechnology* 77, 204–208.

Gamborg, O., Murashige, T., Thorpe, T. and Vasil, I. (1976). Plant tissue culture media. *In Vitro* 12, 473–478.

Gao, J., Yang, F., Zhang, S., Li, J., Chen, H. *et al.* (2014) Expression of a hevein-like gene in transgenic *Agave* hybrid No. 11648 enhances tolerance against zebra stripe disease. *Plant Cell, Tissue and Organ Culture* 119, 579–585.

Griffiths, H. and Males, J. (2017) Succulent plants. *Current Biology* 27, R890–R896.

Gunasena, M., Alahakoon, A., Polwaththa, K., Galpaya, G., Priyanjani, H., Koswattage, K. and Senarath, W. (2024) Transforming plant tissue culture with nanoparticles: a review of current applications. *Plant Nano Biology* 100102. DOI: 10.1016/j.plana.2024.100102

Gupta, A., Naorem, A., Prasad, M.G. and Khuraijam, J.S. (2024) Using wild plant species in ornamental horticulture: a possible future. In: Bhargava, B., Kumar, P., Verma, V. (eds) *Ornamental Horticulture: Latest Cultivation Practices and Breeding Technologies*. Springer, Singapore, pp. 259–273. DOI: 10.1007/978-981-97-4028-4_14

Gutiérrez-Mora, A., Rodríguez-Garay, B., Contreras-Ramos, S.M., Kirchmayr, M.R. and González-Ávila, M. (2014) Sustainable and integral exploitation of *Agave*. CONACYT and CIATEJ, Guadalajara, Mexico. https://www.ciatej.mx/agave/1.7agave.pdf (accessed 10 July 2025).

Hamdeni, I., Yangui, I., Sanaa, A., Slim, S., Louhaichi, M., Messaoud, C., Boulila, A., and Bettaieb, T. (2022) *Aloe vera* L. (Asphodelaceae): supplementation of *in vitro* culture medium with *Aloe vera* gel for production of genetically stable plants. *South African Journal of Botany* 147, 1206–1213.

He, C., Zhang, J., Chen, J., Ye, X., Du, L., Dong, Y. and Zhao, H. (2007) Genetic transformation of *Aloe barbadensis* Miller by Agrobacterium tumefaciens. *Journal of Genetics and Genomics* 34, 1053–1060.

He, W., Wang, D., Yang, N., Cao, D., Chen, X., Chen, J. and Wei, X. (2021). *In vitro* shoot culture of *Sesuvium portulacastrum*: an important plant for phytoremediation. *Agriculture* 12, 47.

Hesami, M. and Jones, A.M.P. (2020) Application of artificial intelligence models and optimization algorithms in plant cell and tissue culture. *Applied Microbiology and Biotechnology* 104, 9449–9485.

Himmelboe, M., Lauridsen, U., Hegelund, J., Müller, R. and Lütken, H. (2013) *Agrobacterium rhizogenes* mediated transformation of *Rhodiola* sp. An approach to enhance the level of bioactive compounds. *Acta Horticulturae* 1098, 144–150.

Hwang, H.-H., Wang, C.-H., Chen, H.-H., Ho, J.-F., Chi, S.-F., Huang, F.-C. and Yen, H. E. (2019) Effective *Agrobacterium*-mediated transformation protocols for callus and roots of halophyte ice plant (*Mesembryanthemum crystallinum*). *Botanical Studies* 60, 1–15.

Ivannikov, R., Lobova, O., Ivannikova, N. and Krasnienkova, I. (2022) micropropagation and organogenesis of *Astrophytum asterias* (Zucc.) Lem .(Cactaceae Juss.), *Blossfeldia liliputiana* Werderm. and *Strombocactus disciformis* (DC.) Britton and Rose. *Journal of Microbiology, Biotechnology and Food Sciences* 11, e2201–e2201.

Kale, M.S., Hake, A. and Thakur, P.P. (2018) *In vitro* regeneration and rapid multiplication of *Kalanchoe blossfeldiana*: an important ornamental plant. *International Journal of Chemistry* 6, 2509–2512.

Kanchanapoom, K., Sunheem, S. and Kanchanapoom, K. (2010) *In vitro* propagation of *Adenium obesum* (Forssk.) Roem. and Schult. *Notulae Botanicae Horti Agrobotanici Cluj-Napoca* 38, 209–213.

Khattab, S. and El Sherif, F. (2011) Effect of growth regulators on *Carpobrotus edulis* rapid micropropagation and molecular analysis. *Journal of American Science* 7, 511–520.

Kim, D.H., Kang, K.W. and Sivanesan, I. (2019) Influence of auxins on somatic embryogenesis in *Haworthia retusa* Duval. *Biologia* 74, 25–33.

Kluge, M. and Ting, I. P. (1978) Crassulacean acid metabolism: analysis of an ecological adaptation (Vol. 30). Springer Science and Business Media, Heidelberg, Germany. DOI: 10.1007/978-3-642-67038-1

Konieczny, R., Obert, B., Bleho, J., Novák, O., Heym, C. *et al.* (2011) Stable transformation of *Mesembryanthemum crystallinum* (L.) with *Agrobacterium rhizogenes* harboring the green fluorescent protein targeted to the endoplasmic reticulum. *Journal of Plant Physiology* 168, 722–729.

Kumar, V. and Jain, M. (2015) The CRISPR–Cas system for plant genome editing: advances and opportunities. *Journal of Experimental Botany* 66, 47–57.

Lecona-Guzmán, C.A., Reyes-Zambrano, S., Barredo-Pool, F.A., Abud-Archila, M., Montes-Molina, J.A., Rincón-Rosales, R. and Gutierrez-Miceli, F.A. (2017) *In vitro* propagation of *Agave americana* by indirect organogenesis. *HortScience* 52, 996–999.

Lema-Rumińska, J. and Kulus, D. (2014) Micropropagation of cacti—a review. *Haseltonia* 2014(19), 46–63.

Leverett, A., Hartzell, S., Winter, K., Garcia, M., Aranda, J. *et al.* (2023) Dissecting succulence: Crassulacean Acid Metabolism and hydraulic capacitance are independent adaptations in *Clusia* leaves. *Plant, Cell and Environment* 46, 1472–1488.

Li, K., Ren, H., Zhao, W., Zhao, X. and Gan, C. (2023) Factors affecting bulblet multiplication in bulbous plants. *Scientia Horticulturae* 312, 111837.

Liu, B., Fang, H., Meng, C., Chen, M., Chai, Q., Zhang, K. and Liu, S. (2017) Establishment of a rapid and efficient micropropagation system for succulent plant *Haworthia turgida* Haw. *HortScience* 52, 1278–1282.

Llamoca-Zárate, R.M., Ponte, L.F.A., Landsmann, J. and Campos, F.d.A.P. (1999) Biolistic-mediated transient gene expression in shoot apical meristems of the prickly-pear (*Opuntia ficus-indica*). *Brazilian Archives of Biology and Technology* 42, 299–302.

Loewus, F.A. and Murthy, P.P. (2000) Myo-inositol metabolism in plants. *Plant Science* 150, 1–19.

Lu, J., Li, S., Deng, S., Wang, M., Wu, Y. *et al.* (2024) A method of genetic transformation and gene editing of succulents without tissue culture. *Plant Biotechnology Journal* 22, 1981–1988.

Lütken, H., Jensen, L. S., Topp, S. H., Mibus, H., Müller, R. and Rasmussen, S. K. (2010) Production of compact plants by overexpression of *AtSHI* in the ornamental *Kalanchoë*. *Plant Biotechnology Journal* 8, 211–222.

Makunga, N., Hall, E. and Stander, M. (2022) Mesembrine alkaloid production in *in vitro* culture morphotypes of *Sceletium tortuosum* (L.) NE Br. *Industrial Crops and Products* 185, 115051.

Martínez, M.I., Barba-Espín, G., Favero, B.T. and Lütken, H. (2020) *Rhizobium rhizogenes*-mediated transformation of *Rhodiola rosea* leaf explants. *Bragantia* 79, 213–223.

Mehbub, H., Akter, A., Akter, M.A., Mandal, M.S.H., Hoque, M.A., Tuleja, M., and Mehraj, H. (2022) Tissue culture in ornamentals: cultivation factors, propagation techniques, and its application. *Plants* 11, 3208.

Miriti, M.N., Rodríguez-Buriticá, S., Wright, S.J. and Howe, H.F. (2007) Episodic death across species of desert shrubs. *Ecology* 88, 32–36.

Mirmazloum, I., Forgács, I., Zok, A., Pedryc, A. and Gyorgy, Z. (2014) Transgenic callus culture establishment, a tool for metabolic engineering of *Rhodiola rosea* L. *Acta Scientiarum Polonorum. Hortorum Cultus* 13(4).

Misra, A.N. and Misra, M. (2012) Chapter 3: Sterilisation techniques in plant tissue culture. In: Sharma, H.D. (ed.) *Plant Tissue Culture: Totipotency to Transgenic.* AGROBIOS, Sardarpura, Jodhpur, Rajasthan, India, pp. 31–41.

Mlungwana, A. (2018) *In vitro* propagation studies of the endangered succulents *Drosanthemum micans* and *Drosanthemum hallii* (Aizoaceae). MSc thesis. Cape Peninsula University of Technology, Cape Town, South Africa. https://etd.cput.ac.za/handle/20.500.11838/2748 (accessed 10 July 2025)

Moebius-Goldammer, K.G., Mata-Rosas, M. and Chávez-Avila, V.M. (2003) Organogenesis and somatic embryogenesis in *Ariocarpus kotschoubeyanus* (Lem.) K. Schum.(Cactaceae), an endemic and endangered Mexican species. *In Vitro Cellular and Developmental Biology-Plant* 39, 388–393.

Mujtaba, M., Wang, D., Carvalho, L.B., Oliveira, J.L., Espirito Santo Pereira, A.d. *et al.* (2021) Nanocarrier-mediated delivery of miRNA, RNAi, and CRISPR-Cas for plant protection: current trends and future directions. *ACS Agricultural Science and Technology* 1, 417–435.

Nakano, M., Nagai, M., Tanaka, S., Nakata, M. and Godo, T. (2005) Adventitious shoot regeneration and micropropagation of the Japanese endangered *Hylotelephium sieboldii* (Sweet ex Hook.) *H. Ohba* and *H. sieboldii* var. *ettyuense* (Tomida) H. Ohba. *Plant Biotechnology* 22, 221–224.

Naz, S., Kayani, H., Saeed, B., Khan, S. and Kauser, N. (2022) An effective protocol for optimization of *in vitro* micropropagation and genetic transformation of *Furcraea foetida*. Research Square, preprint DOI: 10.21203/rs.3.rs-1290605/v1

Nazir, M., Sadat, S. and Soltani Howyzeh, M. (2019) The effect of different hormone combinations on direct and indirect somatic embryogenesis in *Agave americana*. *Iranian Journal of Plant Physiology* 9, 2739–2747.

Newland, K.C., Ives, S., Joseph, G.E., Dimmitt, M.A., Mittleman, M. *et al.* (1981) Propagation techniques for desert plants. I. A minisymposium. *Desert Plants* 2, 205–216.

Niguse, M., Sbhatu, D.B. and Abraha, H.B. (2020) *In vitro* micropropagation of *Aloe adigratana* Reynolds using offshoot cuttings. *The Scientific World Journal* 1, 9645316. DOI:10.1155/2020/9645316

Nobel, P.S. (2003). Environmental biology of agaves and cacti. Cambridge University Press, Cambridge, United Kingdom

Nyffeler, R. and Eggli, U. (2010) An up-to-date familial and suprafamilial classification of succulent plants. *Bradleya* 28, 125–144.

Ofisi, M. (2017) *In vitro* propagation studies of rare *Argyroderma* species strictly endemic to the Knersvlakte region of South Africa. MSc thesis. Cape Peninsula University of Technology, Cape Town, South Africa.

https://etd.cput.ac.za/bitstream/20.500.11838/2714/1/210100982-Ofisi-Mbulelo-MHort-Horticulture-Appsc-2018.pdf (accessed 10 July 2025)

Ong, A.K.S., Prasetyo, Y.T., De Leon, L.A.S., Ayuwati, I.D., Nadlifatin, R. and Persada, S.F. (2022) Plantitas/Plantitos preference analysis on succulents attributes and its market segmentation: Integrating conjoint analysis and K-means clustering for gardening marketing strategy. *Sustainability* 14, 16718.

Orlóci, L. and Fekete, A. (2023) Ornamental plants and urban gardening. *Plants* 12, 4096. DOI: 10.3390/plants12244096

Pandey, S. and Pandey, V. (2018) Plant tissue culture techniques and its achievements. In: Bharadwaj, D.N. (ed.) *Advanced Molecular Plant Breeding: Meeting the Challenge of Food Security*. CRC Press, Boca Raton, FL, United States, pp. 329–366.

Pierik, R.L.M. (1997) *In vitro* culture of higher plants. Martinus Nijhoff publishing, Dordrecht, Netherlands.

Powers, D.E. and Backhaus, R. A. (1989) *In vitro* propagation of *Agave arizonica* Gentry & Weber. *Plant Cell, Tissue and Organ Culture* 16, 57–60.

Rasad, F., Hasbullah, N., Daud, N., Azis, N., Amin, M. and Lassim, M. (2015) Micropropagation of *Adenium obesum* (Dessert Rose) *in vitro*. Proceedings of the International Conference on Agricultural, Ecological and Medical Sciences, International Institute of Chemical, Biological and Environmental Engineering (ICBEE), Phuket, Thailand, pp. 10–12.

Reyes-Díaz, J.I., Arzate-Fernández, A.M., Piña-Escutia, J.L. and Vázquez-García, L.M. (2017) Media culture factors affecting somatic embryogenesis in *Agave angustifolia* Haw. *Industrial Crops and Products* 108, 81–85.

Roberto, S.R. and Colombo, R.C. (2020) Innovation in propagation of fruit, vegetable and ornamental plants. *Horticulturae* 6, 23.

Santacruz-Ruvalcaba, F. and Portillo, L. (2009) Thin cell suspension layer as a new methodology for somatic embryogenesis in *Agave tequilana* Weber cultivar azul. *Industrial Crops and Products* 29, 609–614.

Schenk, R.U. and Hildebrandt, A. (1972) Medium and techniques for induction and growth of monocotyledonous and dicotyledonous plant cell cultures. *Canadian Journal of Botany* 50, 199–204.

Schulte, P.J. and Nobel, P.S. (1989) Responses of a CAM plant to drought and rainfall: capacitance and osmotic pressure influences on water movement. *Journal of Experimental Botany* 61–70.

Seol, E., Jung, Y., Lee, J., Cho, C., Kim, T., Rhee, Y. and Lee, S. (2008) *In planta* transformation of *Notocactus scopa* cv. Soonjung by *Agrobacterium tumefaciens*. *Plant Cell Reports* 27, 1197–1206.

Silos-Espino, H., Valdez-Ortiz, A., Rascón-Cruz, Q., Rodríguez-Salazar, E. and Paredes-López, O. (2006) Genetic transformation of prickly-pear cactus (*Opuntia ficus-indica*) by *Agrobacterium tumefaciens*. *Plant Cell, Tissue and Organ Culture* 86, 397–403.

Stambouli-Essassi, S., Zakraoui, M., Bouzid, S. and Harzallah-Skhiri, F. (2017) Sexual Propagation of the Tunisian spinescent *Opuntia ficus-indica* (L.) Mill., morphogenetic deployment and polymorphism. *Notulae Botanicae Horti Agrobotanici Cluj-Napoca* 45, 50–58.

Tejavathi, D., Rajanna, M., Sowmya, R. and Gayathramma, K. (2007) Induction of somatic embryos from cultures of *Agave vera-cruz* Mill. *In Vitro Cellular and Developmental Biology-Plant* 43, 423–428.

Thirukkumarana, G., Ntui, V. O., Khan, R. S., Nakamura, I. and Mii, M. (2010) Generation of phenotypically normal marker-free transgenic plants of *Kalanchoe blossfeldiana* through hairy root induction. *Plant Biotechnology* 27, 147–153.

Wyka, T., Wróblewska, M. and Hamerska, M. (2009) Use of cactus flowers as explants for micropropagation. *The Journal of Horticultural Science and Biotechnology* 84, 454–458.

Tung, H.T., Bao, H.G., Cuong, D.M., Ngan, H.T.M., Hien, V.T. *et al.* (2021a) Silver nanoparticles as the sterilant in large-scale micropropagation of chrysanthemum. *In Vitro Cellular and Developmental Biology-Plant* 57, 897–906. DOI: 10.1007/s11627-021-10163-7

Tung, H.T., Thuong, T.T., Cuong, D.M., Luan, V.Q., Hien, V.T. *et al.* (2021b) Silver nanoparticles improved explant disinfection, *in vitro* growth, runner formation and limited ethylene accumulation during micropropagation of strawberry (*Fragariax ananassa*). *Plant Cell, Tissue and Organ Culture* 145, 393–403.

Valenzuela-Sánchez, K.K., Juárez-Hernández, R.E., Cruz-Hernández, A., Olalde-Portugal, V., Valverde, M.E. and Paredes-Lopez, O. (2006) Plant regeneration of *Agave tequilana* by indirect organogenesis. *In Vitro Cellular and Developmental Biology-Plant* 42, 336–340.

Von Willert, D., Eller, B., Werger, M. and Brinckmann, E. (1990) Desert succulents and their life strategies. *Vegetatio* 90, 133–143.

Wang, X., Chen, X., Cheng, Q., Zhu, K., Yang, X. and Cheng, Z. (2019). *Agrobacterium*–mediated Transformation of *Kalanchoe laxiflora*. *Horticultural Plant Journal* 5, 221–228.

Wijerathna-Yapa, A. and Hiti-Bandaralage, J. (2023) Tissue culture—a sustainable approach to explore plant stresses. *Life* 13, 780.

Winter, K. and Smith, J.A.C. (2022) CAM photosynthesis: the acid test. *New Phytologist* 233, 599–609.

Zhang, Z., Tang, W., Wang, Y., Wang, Y., Sun, M. *et al.* (2021) Explant disinfection with chlorine dioxide promotes adventitious shoot organogenesis in *Crassula portulacea* via the regulation of reactive oxygen species and related gene expression. *Plant Cell, Tissue and Organ Culture* 146, 531–540.

Zhou, L., Yang, Y., Sharma, A., Singh, V. P., Tripathi, D. K. *et al.* (2024) A rapid method for obtaining the transgenic roots of Crassulaceae plants. *Plants* 13, 3024.

8 Applications and Future Perspectives of Genome-Wide Association Studies in Horticultural Flower Crops

Chong-Wei Lee and Shu-Yun Chen*
Department of Agronomy, National Chung-Hsing University, Taichung, Taiwan

Abstract

Horticultural crops, including both ornamental and non-ornamental species, possess significant economic value due to their aesthetic and edible properties. Researchers have been dedicated to studying various traits of horticultural crops to enhance their commercial value. In particular, understanding the genetic basis of these traits has become a major focus, driving the application of genome-wide association studies (GWAS) in horticultural research. GWAS has been widely employed to uncover the genetic mechanisms underlying trait variation, offering valuable insights for trait improvement and molecular breeding. This review highlights key successful GWAS applications in horticultural flower crops, such as orchids, roses, chrysanthemums, and sunflowers, discusses the challenges associated with its implementation, and explores possibilities for advancing GWAS in horticulture. By providing a comprehensive overview, this review aims to deepen the understanding of GWAS as a powerful genomic tool for horticultural research.

Keywords: Horticultural crops, ornamental crops, sequencing, GWAS, trait selection

8.1 Introduction

8.1.1 The evolution of sequencing technology and the origin and development of genome-wide association studies (GWAS)

8.1.1.1 The development of sequencing technology

The introduction of Sanger sequencing in 1977 marked a key moment in genomics, laying the foundation for DNA sequencing. While revolutionary, its high cost and limited efficiency made it difficult to perform large-scale genomic studies. Technological advances over the following decades, however, brought forth next-generation sequencing (NGS), fundamentally transforming the field of genomics. Platforms such as Illumina and PacBio significantly reduced sequencing costs and turnaround time while offering deeper coverage, higher throughput, and improved accuracy (Venter *et al.*, 2001; Koboldt *et al.*, 2013; Korte and Farlow, 2013; Tam *et al.*, 2019). The sequencing

*Corresponding author: syc@nchu.edu.tw and ubs717@gmail.com

 Biotechnology of Orchids and Other Ornamentals (Eds P. Bolaños-Villegas and F-C. Chen)
DOI: 10.1079/9781800627895.0008

cost has significantly declined over the years, with data sourced from the National Human Genome Research Institute (Wetterstrand, n.d.). We have organized and presented this data in Figure 8.1.

Next generation sequencing has become an indispensable tool for modern genomics, enabling rapid, high-resolution sequencing on an unprecedented scale. It has propelled the assembly of genome-wide datasets across diverse species, including model plants like *Arabidopsis*, tomato, and tobacco, as well as staple crops such as rice and wheat (Desaint *et al.*, 2023). Additionally, genomic sequencing data for economically significant crops like sweet potato, cotton, and coffee have been made available, enriching genomic databases and providing researchers with the resources necessary to explore genetic variation and complex traits (Koboldt *et al.*, 2013; Desaint *et al.*, 2023).

In recent years, the rise of third-generation sequencing technologies—notably Oxford Nanopore—has introduced further innovation. These platforms offer unique advantages, such as long-read capabilities, portability, and cost-effectiveness, facilitating the assembly of complex genomes with greater precision (Lu *et al.*, 2016; Uffelmann *et al.*, 2021). The integration of long reads from third-generation sequencing with short reads from NGS represents a powerful hybrid approach, enhancing sequencing accuracy and delivering more comprehensive genomic assemblies (Lu *et al.*, 2016; Koboldt *et al.*, 2013; Desaint *et al.*, 2023).

The continuous evolution of sequencing technologies has dramatically reduced barriers to genomic research. Today, genomic data are more abundant, accurate, and representative of a diverse array of species. These advances serve as a cornerstone for genome-wide association studies (GWAS), enabling researchers to decode the genetic basis of complex traits and further accelerating breakthroughs in fields like crop genomics and molecular breeding.

8.1.1.2 Origin and development of genome-wide association studies (GWAS)

GWAS represent a transformative milestone in genomics, enabling the identification of genetic variations associated with complex traits, such as yield, disease resistance, and quality, across horticultural and agricultural crops (Tibbs Cortes *et al.*, 2021; Desaint *et al.*, 2023). This

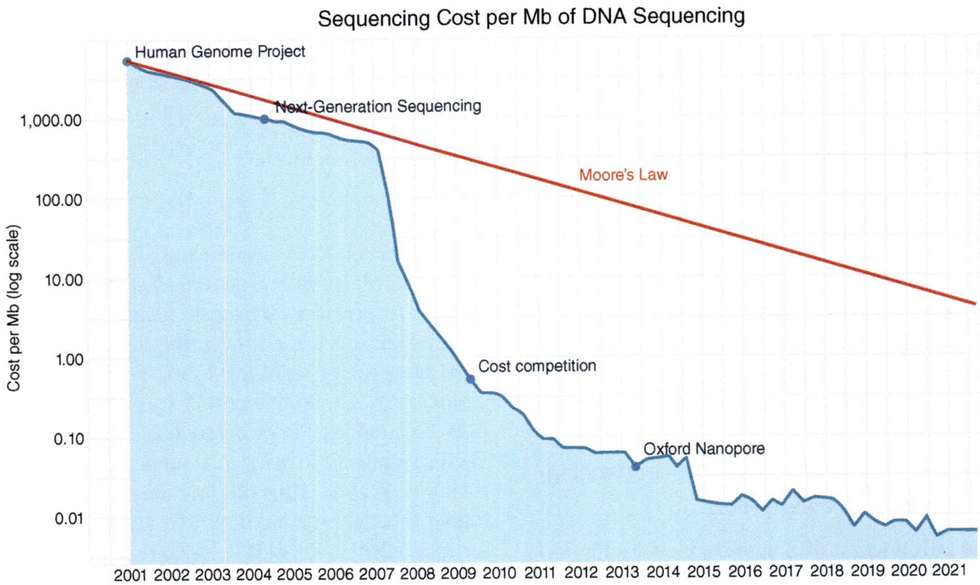

Fig. 8.1. Accelerated reduction in the cost of sequencing cost per megabase (log scale) due to technological breakthroughs has outpaced Moore´s Law, an observation that postulates that the computing capacity of a system doubles every two years. This figure is original, from data sourced and adapted from the National Human Genome Research Institute (Wetterstrand, n.d.).

approach leverages natural genetic diversity, combined with large-scale phenotypic and genotypic data, to unravel the intricate genetic architecture underlying economically important traits.

The quick evolution of sequencing technologies, particularly the advent of high-throughput platforms, has catalyzed the widespread adoption of GWAS. Lower sequencing costs and the availability of extensive genomic databases for key crops—ranging from staple grains to high-value horticultural species—have established a robust foundation for GWAS-based investigations (Tam *et al.*, 2019). This method's greatest strength lies in its ability to statistically associate single-nucleotide polymorphisms (SNPs) across entire genomes with phenotypic variation across natural populations (Gallagher and Chen-Plotkin, 2018).

The necessity for GWAS arose from the limitations of traditional genetic approaches. Traits such as yield, quality, and disease resistance in crops are inherently polygenic, and influenced by many loci and by environmental interactions (Korte and Farlow, 2013). Early genetic studies often relied on biparental quantitative trait loci (QTL) mapping, that provided some insight but were limited by little homologous recombination and by low-resolution mapping. GWAS, in contrast, exploits the genetic diversity already present in natural or germplasm populations, which may undergo multiple recombination events over generations. This allows for higher mapping resolution and enables the precise localization of causal loci at the polymorphism level (Desaint *et al.*, 2023).

GWAS studies require the integration of massive genomic and phenotypic datasets, necessitating advanced statistical approaches to account for confounding factors like population structure, kinship, and gender. Statistical frameworks, such as general linear models (GLM) and mixed linear models (MLMs), have been developed to address biases and improve the robustness of GWAS findings. These models, alongside emerging species-specific tools, have elevated GWAS into a premier method for dissecting polygenic traits (Tibbs Cortes *et al.*, 2021).

Initially pioneered in human genetics for disease susceptibility studies, GWAS has found widespread application in plant and animal genomics. For crops like rice, wheat, and maize, GWAS has been instrumental in identifying genes associated with disease resistance, productivity, and quality traits. These findings have profound implications for molecular breeding. Techniques such as marker-assisted selection (MAS) enable breeders to expedite the development of superior varieties by directly targeting identified loci, enhancing crop resilience, yield, and market value (Visscher *et al.*, 2012; Gallagher and Chen-Plotkin, 2018).

Recent advancements in data analytics and sequencing platforms continue to expand GWAS applications. The integration of long-read sequencing technologies and improved statistical tools promises even greater precision in linking genotype to phenotype, particularly in species with complex polyploid genomes. As genomic resources for horticultural crops continue to grow, GWAS will undoubtedly play a significant role in addressing global agricultural challenges. By enhancing our understanding of crop genetics, this powerful tool will drive innovation in sustainable agriculture and bolster food security worldwide (Uffelmann *et al.*, 2021; Desaint *et al.*, 2023).

8.1.2 Introduction to horticultural crops

8.1.2.1 High economic and added value

Horticultural crops, encompassing fruits, vegetables, flowers, and ornamental plants, are critical contributors to the global economy. These crops not only serve as vital sources of nutrition but also deliver substantial economic value due to their diverse applications and high market demand. Rich in essential nutrients, such as vitamins, minerals, and antioxidants, horticultural crops address consumer preferences for fresh, high-quality produce while promoting public health (Slavin and Lloyd, 2012).

Beyond their nutritional importance, floricultural crops—such as roses, tulips, and orchids—hold significant aesthetic and economic value. They are widely utilized in decorative applications, including weddings, cultural events, and landscape design, where their vibrant colors and exquisite traits enhance market competitiveness (Zheng *et al.*, 2021). The high economic returns associated with horticultural crops are intrinsically linked to their genomic complexity. For instance, many horticultural species show polyploidy—having multiple sets of chromosomes—which amplifies their genetic variability. This genomic characteristic often manifests as larger fruit size, improved nutritional content, and more visually appealing flowers, further increasing the added value of these crops (Sattler *et al.*, 2016; Van de Peer *et al.*, 2017).

Thus, the economic importance of horticultural crops is not only driven by market demand but is also fueled by their genetic diversity, which creates opportunities for breeding innovations to enhance crop traits.

8.1.2.2 Characteristics and genomic complexity of horticultural crops

Horticultural crops are characterized by exceptional genetic diversity and complex genomic architectures. Many economically significant species, such as strawberries, blueberries, and orchids, are polyploids with multiple sets of chromosomes. Polyploidy provides an evolutionary advantage by supplying great genetic variation, enabling these crops to express a diverse array of traits. For instance, polyploidy in strawberries enhances their fruit sweetness and quality, while in blueberries, it contributes to improved disease resistance and adaptability to various climatic conditions (Colle *et al.*, 2019; Edger *et al.*, 2019). Similarly, the polyploid genome of orchids supports the production of vibrant, long-lasting flowers, enhancing their ornamental appeal and value in the global market (Zhang *et al.*, 2017).

While polyploidy enriches genetic variability, it also introduces challenges for genomic research and breeding. The increased gene copy number, interactions between homologous chromosomes, and the complex regulation of gene expression result in phenotypes that cannot be attributed to single genes alone (Osborn *et al.*, 2003). This genomic heterogeneity complicates the identification of causal genetic variations. However, this complexity also provides horticultural crops with significant phenotypic plasticity. Such adaptability enables crops to respond to environmental changes, ensuring their survival and performance across diverse growing conditions (Soltis and Soltis, 2000; Chen, 2007).

Traditional genetic approaches, such as QTL analysis, have limitations in addressing the genomic complexity of horticultural crops. QTL mapping, which relies on biparental populations, often identifies broad genomic regions with limited resolution. This reduces the ability to pinpoint specific genetic variations by controlling target traits. In contrast, GWAS offers a more refined approach. By analyzing natural populations with diverse genetic backgrounds, GWAS leverages historical recombination events to achieve higher-resolution mapping of trait-associated loci (Tam *et al.*, 2019).

The power of GWAS lies in its ability to simultaneously scan tens of thousands of SNP markers across entire genomes and to link them to existing phenotypic variation. This capability allows researchers to identify specific loci associated with complex traits of horticultural interest. Coupled with advancements in high-throughput sequencing, GWAS has emerged as an indispensable tool for dissecting polygenic traits and accelerating molecular breeding efforts (Desaint *et al.*, 2023).

By integrating large-scale genomic and phenotypic data, GWAS enables breeders to harness the genetic potential of horticultural crops, enhancing their productivity, resilience, and market value. In this way, the genomic complexity of horticultural crops—once a research challenge—now represents an opportunity for transformative advancements in molecular breeding and sustainable agriculture.

8.2 Research Applications, Barriers, and Challenges of GWAS in Horticultural Crops

Since its arrival, GWAS has emerged as a remarkable tool for decoding the genetic architecture of horticultural crops. Enabled by rapid advancements in sequencing technologies and the exponential growth of genomic data, GWAS now allows researchers to pinpoint genetic variations associated with key traits such as yield, quality, stress resistance, and morphological attributes. This transformative approach has not only accelerated molecular breeding programs but also shed light on the underlying genetic mechanisms of complex traits in horticultural crops.

8.2.1 Case studies of non-floral horticultural crops

Case 1: Cold tolerance in tomatoes

Tomato (*Solanum lycopersicum*) is one of the most widely cultivated crops globally, yet its sensitivity to low temperatures limits its production in colder regions. Zhu *et al.* (2023) from Henan University, in Kaifeng, China employed GWAS to uncover genetic determinants of cold tolerance and identified a key variation in the B-box zinc-finger (BBX) 31 (*SlBBX31*) gene promoter that causes loss of function (Table 8.1). This variation modulates the expression of multiple cold-responsive genes, enhancing cold tolerance in tomatoes. By elucidating these gene expression networks, the study provides actionable genetic markers for breeding climate-resilient tomato varieties, improving their adaptability to challenging environments.

Table 8.1. Summary of GWAS case studies in high profile horticultural crops. Author's own table.

Category	Case title	Research findings	Reference
Non-floral horticultural crops	Cold tolerance in tomatoes	Identified variation in the *SlBBX31* promoter modulating cold-responsive genes for enhanced cold tolerance	Zhu *et al.*, 2023
	Powdery mildew resistance in cucumbers	Identified SNPs on chromosome 5 and candidate genes *Csa5G453160* and *Csa5G471070* linked to resistance	Lee *et al.*, 2020
	Sex expression in oriental melon	Identified 18 SNPs and four candidate genes regulating sex expression and fruit morphology	Kishor *et al.*, 2021
	Fruit quality of sweet cherry and strawberry	Identified SNPs related to fruit quality traits and 39 QTLs for strawberry attributes	Donkpegan *et al.*, 2023; Rehman *et al.*, 2024
	Fruit size and shape in apples	Identified 59 SNPs and candidate genes *MdOFP17* and *MdOFP4* regulating fruit shape	Dujak *et al.*, 2024
	Fusarium wilt resistance in *Luffa*	Identified 6 QTLs and a marker linked to AUDPC with 93.75% accuracy	Li *et al.*, 2024
Floral horticultural crops	Rose petal color	Identified five loci for anthocyanins and two loci for carotenoids influencing petal color	Schulz *et al.*, 2016
	Pigment variation in sunflowers	Identified *HaMYB111* regulating UV-absorbing pigments for pollination and drought tolerance	Todesco *et al.*, 2022
	Genomics of rose flower traits	Mapped loci controlling petal number, flower size, and flowering time	Saint-Oyant *et al.*, 2018
	Adventitious root formation in roses	Identified *SAC9* gene critical for root development and auxin transport	Wamhoff *et al.*, 2023
	Flower morphological traits in Chrysanthemums	Identified QTLs affecting floral structure and ornamental value	Song *et al.*, 2020
	Genetic architecture of plum blossom flower morphology	Identified loci influencing petal number and flower size	Zheng *et al.*, 2021
	Flower traits in Cymbidium orchids	Identified QTLs linked to anthocyanin synthesis for vibrant colors	Hsu *et al.*, 2022
	Multi-petal traits in orchids	Found RBE gene controlling *CsAP3* and *CsPI* for petal traits	Yang *et al.*, 2023
	Leaf angle in chrysanthemums	Identified 303 QTLs and genes *ARF16*, *IBH1*, and *AN3* regulating leaf angle	Lou *et al.*, 2023

Case 2: Powdery mildew resistance in cucumbers

Cucumber (*Cucumis sativus*), an economically significant crop, frequently suffers from powdery mildew, which compromises its yield and quality. Lee *et al.* (2020) from Sejong University and Seoul National University, in Korea, utilized GWAS on a global collection of 264 cucumber accessions to identify SNPs linked to powdery mildew resistance (Table 8.1). Significant associations were mapped to chromosome 5, where two candidate genes, *Csa5G453160* and *Csa5G471070*, were implicated. These genes regulate resistance mechanisms, offering critical molecular targets for MAS to develop powdery mildew-resistant cucumber cultivars.

Case 3: Sex expression in oriental melon

The oriental melon (*Cucumis melo* var. *makuwa*), a staple in East Asian agriculture, exhibits sex-specific traits that influence yield and fruit morphology. Kishor *et al.* (2021) from Dongguk University, in Korea, conducted a GWAS analysis across 48 melon varieties, that identified 18 SNPs linked to fruit morphology and sex expression (Table 8.1). Two novel SNPs on chromosomes 1 and 8, along with four candidate genes (*MELO3C015898*, *MELO3C015904*, *MELO3C024563*, and *MELO3C024565*), were identified as key regulators of sex expression. These findings provide molecular insights into floral development and offer a foundation for breeding melons with optimized yield and fruit quality, including traits such as smooth surface and yellow to orange color.

Case 4: Fruit quality of sweet cherry and strawberry

Sweet cherry (*Prunus avium*) fruit quality traits—including size, firmness, and crack resistance—are critical for market acceptance. Donkpegan *et al.* (2023) from the University of Bordeaux, in France, employed multi-year phenotyping and GWAS analysis of 116 cherry varieties, identifying SNPs associated with these traits (Table 8.1). Candidate genes primarily involved in hormone regulation and cell wall metabolism were highlighted, offering new avenues for improving cherry fruit quality through MAS.

For strawberry (*Fragaria* × *ananassa*), consumer demand hinges on attributes such as sweetness, firmness, and acidity. Rehman *et al.* (2024) identified 39 QTLs across the strawberry genome associated with fruit quality, winter survival, and flowering time (Table 8.1). By pinpointing genomic regions regulating fruit texture and sweetness, this study enables precision breeding strategies to develop strawberries with superior market appeal.

Case 5: Fruit size and shape in apples

Apple (*Malus domestica*), a cornerstone of global horticulture, relies heavily on fruit size and shape for its economic value. Dujak *et al.* (2024) from the Centre for Research in Agricultural Genomics (CRAG) in Barcelona, Spain, applied GWAS using a high-density SNP array across 355 apple genotypes (Table 8.1). They identified 59 SNPs that influence fruit morphology and were distributed across all chromosomes except for chromosomes 10 and 15. Notably, candidate genes *Ovate Family Proteins 17/4* (*MdOFP17* and *MdOFP4*)—members of the *OVATE* family—were linked to fruit shape regulation. These findings provide essential molecular markers for apple breeding programs, facilitating the development of varieties with improved aesthetic and market traits.

Case 6: Fusarium wilt resistance in Luffa

In line with our laboratory's research focus, Li *et al.* (2024) conducted a single-plant GWAS (sp-GWAS) study to explore *Fusarium* wilt resistance in 97 individuals from 73 *Luffa* accessions (Table 8.1). Through double digest restriction-site associated DNA (ddRAD) sequencing, the study identified 8919 high-quality SNPs and six QTLs associated with key traits, including Area Under the Disease Progress Curve (AUDPC). Notably, a QTL on chromosome 2 was found to contain the candidate gene *LOC111009722*, which encodes peroxidase 40—a protein critical for plant disease resistance.

Furthermore, this research team developed and validated an SNP marker linked to AUDPC with a 93.75% accuracy in identifying susceptible individuals. This marker provides a valuable tool for MAS, enabling the breeding of *Fusarium* wilt-resistant *luffa* varieties. The study highlights the efficiency of sp-GWAS in identifying actionable genetic markers for enhancing crop resilience (Li *et al.*, 2024).

8.2.2 Case studies of floral horticultural crops

8.2.2.1 Color traits

CASE 1: ROSE PETAL COLOR. Schulz *et al.* (2016) from the Institute for Plant Genetics, in Leibnitz University, Germany, conducted a GWAS analysis on 96 tetraploid rose cultivars uncovering the genetic basis of anthocyanin and carotenoid content (Table 8.1). The study identified five genetic regions influencing anthocyanin levels, which impart red pigmentation, and two loci associated with carotenoid accumulation, resulting in yellow and orange hues. These findings provide valuable molecular markers for enhancing color stability and expanding market appeal in rose breeding.

CASE 2: PIGMENT VARIATION IN SUNFLOWERS. Todesco *et al.* (2022) from the University of British Columbia, Canada, explored the relative abundance of ultraviolet (UV)-absorbing pigments in sunflowers, focusing on their genetic regulation (Table 8.1). By means of GWAS they identified the gene *HaMYB111* as a key regulatory gene controlling UV absorption patterns by controlling the synthesis and accumulation of quercetin (a flavonol) and di-*O*-caffeoyl quinic acid. These pigments also enhance environmental resilience by minimizing water loss under drought stress. This study provides a foundation for improving both aesthetic appeal and drought tolerance in sunflowers.

8.2.2.2 Morphology and flower development

CASE 1: GENOMICS OF ROSE FLOWER TRAITS. Hibrand Saint-Oyant *et al.* (2018) from the INRA and the Université d'Angers in France, developed a high-quality genomic map of allopolyploid rose variety 'Old Blush' and used GWAS to pinpoint genetic regions controlling flower traits (Table 8.1). Key loci were discovered for petal number, apparently controlled by a rose *APETALA2/TOE* homolog, flower size, and flowering time, facilitating the development of novel rose varieties with improved ornamental and economic value.

CASE 2: ADVENTITIOUS ROOT FORMATION IN ROSES. Wamhoff *et al.* (2023) from Leibniz Universität, in Hannover, Germany, investigated adventitious root formation in 95 rose genotypes (*Rosa chinensis)* using GWAS (Table 8.1). This is a key trait for the successful propagation by cuttings that are grafted onto rootstocks. The *SAC9* gene was identified as critical for root system development, playing roles in endocytosis and auxin transport and homeostasis by the PIN1 transporter. These findings highlight a genetic basis for improving propagation efficiency in rose breeding programs.

CASE 3: FLOWER MORPHOLOGICAL TRAITS IN CHRYSANTHEMUMS. Song *et al.* (2020) analyzed the number and arrangement of tubular and ligulate florets in chrysanthemums using GWAS (Table 8.1). The study identified several QTLs affecting floral structure, which have a notable impact on the ornamental characteristics of chrysanthemums. These genetic insights enable breeders to select for market-preferred floral forms more efficiently.

CASE 4: GENETIC ARCHITECTURE OF PLUM BLOSSOM FLOWER MORPHOLOGY. Zheng *et al.* (2021) conducted GWAS to identify loci influencing petal number and flower size in plum blossoms (Table 8.1). These loci regulate floral structure, enhancing ornamental appeal while also providing genetic opportunities for precise breeding of plum varieties.

CASE 5: FLOWER TRAITS IN PLUM BLOSSOMS. Li *et al.* (2022) conducted a comprehensive GWAS on flower size, shape, and color in plum blossoms (Table 8.1). Key loci located on multiple chromosomes were associated with floral morphology and plant growth, offering tools to enhance phenotypic traits and competitiveness in the horticultural market

CASE 6: FLOWER TRAITS IN ORCHIDS. Hsu *et al.* (2022) applied GWAS to analyze petal color and size in *Phalaenopsis* orchids (Table 8.1). Several QTLs linked to anthocyanin synthesis were identified, influencing vibrant colors like red and yellow spots. These findings provide a basis for molecular breeding to develop more visually appealing orchid varieties.

8.2.2.3 Specialized traits

CASE 1: MULTI-PETAL TRAITS IN ORCHIDS. Yang *et al.* (2023) studied 195 Chinese orchid cultivars to identify genetic regulators of multi-petal traits (Table 8.1). The *RABBIT EARS* (*RBE*) gene was found to control *C. sinense APETALA3* (*CsAP3*) and *C. sinense PISTILLATA* (*CsPI*), two critical genes governing petal number and shape. Overexpression experiments confirmed the role of *RBE*, offering potential for breeding multi-petal orchids with enhanced aesthetic value.

CASE 2: LEAF ANGLE IN CHRYSANTHEMUMS. Lou *et al.* (2023) performed GWAS on 200 chrysanthemum lines to investigate leaf angle, a trait critical for photosynthetic efficiency and plant architecture (Table 8.1). Using 330,710 SNP markers, the study identified 303 QTLs and seventy-five loci interacting with environmental factors. Genes such as *auxin response factor 16* (*ARF16*), *ILI1 binding bHLH 1* (*IBH1*), and *ANGUSTIFOLIA3* (*AN3*) were highlighted as regulators, offering strategies to optimize plant structure for increased productivity and market value.

8.3 Obstacles and Challenges in GWAS Applications

While GWAS have significantly advanced horticultural crop research, their implementation continues to face many hurdles.

8.3.1 High genomic complexity of polyploid crops

Polyploid horticultural crops, such as strawberries and orchids, present substantial challenges for GWAS analysis due to their overly complex genomic structures. These crops possess multiple sets of chromosomes, leading to gene redundancy and the presence of multiple gene copies. This complexity obscures the identification of genetic variations associated with traits, as researchers must manage mixed signals from overlapping genetic loci. Advanced bioinformatics tools and polyploid-specific statistical models are essential to disentangle these variations and improve the accuracy of genotype-to-phenotype associations.

8.3.2 The role of environmental influence on trait expression

Horticultural crop traits are inherently sensitive to environmental factors such as temperature, light intensity, and water availability. These variables significantly affect the phenotypic expression of key traits like flower color, fruit flavor, and disease resistance, complicating the reproducibility of GWAS findings. To mitigate these inconsistencies, researchers must perform multi-environment trials and validate results across diverse climatic conditions. Integrating environmental interaction models into GWAS frameworks will further strengthen the robustness of trait associations.

8.3.3 Population structure and false positives

The genetic diversity of horticultural crops, shaped by generations of breeding and selection, often results in heterogeneous population structures. This heterogeneity can introduce false positives or spurious associations in GWAS analyses. To account for such biases, researchers commonly utilize statistical approaches like MLM, which adjust for population structure and kinship effects. However, the development of more sophisticated models capable of accounting for complex genomic relationships remains critical (Demirjian *et al.*, 2023).

8.3.4 The problem of having limited genomic resources

For certain economically valuable horticultural crops, especially those with limited research investments, genomic resources remain insufficient. The absence of high-quality reference genomes and comprehensive genotypic datasets restricts the power of GWAS to uncover meaningful associations. Expanding sequencing efforts and establishing publicly accessible genomic databases will be instrumental in unlocking the genetic potential of under-researched species.

8.4 Prospects for the Future

8.4.1 Technological advancements in sequencing

The continuous evolution of high-throughput sequencing technologies promises to revolutionize GWAS applications in horticultural crops. Declining sequencing costs and faster data generation will enable researchers to analyze large-scale genotypic and phenotypic datasets with unprecedented resolution. These advancements are particularly relevant for polyploid crops, where accurate mapping of duplicated chromosomes and multi-gene copies has remained a bottleneck. Progress in polyploid genomics will allow for precise dissection of genetic loci underlying complex traits.

8.4.2 Emergence of third-generation sequencing technologies

Third-generation technologies, such as Oxford Nanopore, offer the ability to read long DNA fragments, facilitating the assembly of complex genomes. Combining long-read sequencing with short-read data from NGS will significantly improve the quality and completeness of horticultural crop genomes. This synergy enhances the accuracy of GWAS results, enabling researchers to pinpoint causal genetic variants for key traits.

8.4.3 Improved statistical models

Advancements in statistical tools are progressively addressing the challenges of environmental variation and population heterogeneity in GWAS. Emerging models that integrate genotype, phenotype, and environmental interaction data promise improved resolution in trait mapping. These innovations aim to minimize false positives and false negatives, enabling GWAS to unravel the genetic architecture of polygenic traits, such as fruit flavor, color stability, and yield.

Increasingly, researchers are leveraging machine learning to enhance GWAS accuracy and expand its analytical capabilities. Machine learning effectively addresses key limitations of traditional statistical approaches. While GWAS may identify several SNP markers associated with key traits, distinguishing causal variants from those linked by linkage disequilibrium remains a significant challenge. Furthermore, GWAS struggles to detect SNPs with small effect sizes and to model complex non-additive interactions such as epistasis. Machine learning, with its ability to manage nonlinear relationships and intricate data patterns, strengthens both the accuracy and interpretability of GWAS results (Nicholls *et al.*, 2020; Sun *et al.*, 2021; Liu *et al.*, 2024; Upadhyaya *et al.*, 2024).

Machine learning contributes to GWAS in several ways:

1. **Prioritization of candidate variants and genes:** techniques like random forests and gradient boosting machines help rank SNPs and genes by their potential causal impact, streamlining downstream analyses. (Nicholls *et al.*, 2020; Sun *et al.*, 2021).
2. **Prediction of complex traits:** by incorporating SNP data, kinship, and environmental factors, machine learning significantly improves phenotype prediction, such as disease resistance in plants, surpassing traditional methods. (Liu *et al.*, 2024; Upadhyaya *et al.*, 2024).
3. **Detection of epistasis:** neural networks and other machine learning models reveal gene–gene interactions that are often missed in single-SNP GWAS analyses (Sun *et al.*, 2021).

In summary, integrating machine learning into GWAS bridges critical gaps by refining variant prioritization, capturing complex genetic architectures, and improving prediction models for functional studies and breeding programs. These advancements promise to unlock deeper insights into the genetic basis of polygenic traits and drive progress in precision agriculture and genomics.

8.4.4 Integration with molecular breeding and gene editing

The insights gained through GWAS will play a pivotal role in the molecular breeding of horticultural crops. Key applications include:

- **Marker-assisted selection (MAS):** breeders can use identified genetic markers to accelerate the selection of desirable traits, such as enhanced **disease resistance**, **improved yields**, and superior **fruit quality**.
- **Gene-editing technologies:** tools like **CRISPR/Cas9** can be applied to modify specific loci identified through GWAS, enabling precise trait improvements with unprecedented efficiency.

8.4.5 Toward innovation and sustainability

The integration of GWAS with emerging technologies, such as **artificial intelligence** and **big data analytics**, will drive innovation in horticultural crop research. Combining genomics with phenomics and environmental data will provide holistic insights into trait regulation, particularly for crops affected by climate change. Such breakthroughs will enhance the **competitiveness** and **sustainability** of the horticultural industry.

8.5 Conclusions

As sequencing technologies advance and our understanding of polyploid crop genomes such as rose deepens, GWAS will continue to transform horticultural crop research. The synergistic application of GWAS, gene editing, and data-driven tools will enable the development of high-value varieties with improved resilience, productivity, and aesthetic appeal. These innovations will not only drive agricultural progress but also contribute to an **enhanced quality of life**, positioning floral horticultural crops as vital components of a sustainable future.

References

Chen, Z.J. (2007) Genetic and epigenetic mechanisms for gene expression and phenotypic variation in plant polyploids. *Annual Reviews on Plant Biology* 58, 377–406.

Colle, M., Leisner, C.P., Wai, C.M., Ou, S., Bird, K.A. *et al.* (2019). Haplotype-phased genome and evolution of phytonutrient pathways of tetraploid blueberry. *GigaScience* 8, giz012.

Demirjian, C., Vailleau, F., Berthomé, R. and Roux, F. (2023) Genome-wide association studies in plant pathosystems: success or failure? *Trends in Plant Science* 28, 471–485.

Desaint, H., Hereil, A. and Causse, M. (2023) Genome-wide association study: A powerful approach to map QTLs in crop plants. In: Raina, A., Wani, M.R., Laskar, R.A., Tomlekova, N. and Khan, S. (eds) *Advanced Crop Improvement, Volume 1: Theory and Practice*. Springer, Cham, Switzerland, pp. 379–455.

Donkpegan, A.S., Bernard, A., Barreneche, T., Quero-García, J., Bonnet, H. *et al.* (2023) Genome-wide association mapping in a sweet cherry germplasm collection (*Prunus avium* L.) reveals candidate genes for fruit quality traits. *Horticulture Research* 10, uhad191.

Dujak, C., Coleto-Alcudia, V. and Aranzana, M.J. (2024) Genomic analysis of fruit size and shape traits in apple: unveiling candidate genes through GWAS analysis. *Horticulture Research* 11, uhad270.

Edger, P.P., Poorten, T.J., VanBuren, R., Hardigan, M.A., Colle, M. *et al.* (2019) Origin and evolution of the octoploid strawberry genome. *Nature Genetics* 51, 541–547.

Gallagher, M.D. and Chen-Plotkin, A.S. (2018) The post-GWAS era: from association to function. *The American Journal of Human Genetics* 102, 717–730.

Hibrand Saint-Oyant, L., Ruttink, T., Hamama, L., Kirov, I., Lakhwani, D., Zhou, N.N. *et al.* (2018) A high-quality genome sequence of *Rosa chinensis* to elucidate ornamental traits. *Nature Plants* 4, 473–484.

Hsu, C.C., Chen, S.Y., Chiu, S.Y., Lai, C.Y., Lai, P.H. *et al.* (2022) High-density genetic map and genome-wide association studies of aesthetic traits in *Phalaenopsis* orchids. *Scientific Reports* 12(1), 3346.

Kishor, D.S., Noh, Y., Song, W.H., Lee, G.P., Park, Y. *et al.* (2021) SNP marker assay and candidate gene identification for sex expression via genotyping-by-sequencing-based genome-wide associations (GWAS) analyses in Oriental melon (*Cucumis melo* L. *var. makuwa*) *Scientia Horticulturae* 276, 109711.

Koboldt, D.C., Steinberg, K.M., Larson, D.E., Wilson, R.K. and Mardis, E.R. (2013) The next-generation sequencing revolution and its impact on genomics. *Cell* 155, 27–38.

Korte, A. and Farlow, A. (2013) The advantages and limitations of trait analysis with GWAS: a review. *Plant Methods* 9, 1-9.

Lee, H.Y., Kim, J.G., Kang, B.C. and Song, K. (2020) Assessment of the genetic diversity of the breeding lines and a genome wide association study of three horticultural traits using worldwide cucumber (*Cucumis* spp.) germplasm collection. *Agronomy* 10, 1736.

Li, M., Sang, M., Wen, Z., Meng, J., Cheng, T. *et al.* (2022) Mapping floral genetic architecture in *Prunus mume*, an ornamental woody plant. *Frontiers in Plant Science* 13, 828579.

Li, Y.D., Liu, Y.C., Jiang, Y.X., Namisy, A., Chung, W.H. *et al.* (2024) Analyzing genetic diversity in *luffa* and developing a *Fusarium* wilt-susceptible linked SNP marker through a single plant genome-wide association (sp-GWAS) study. *BMC Plant Biology* 24, 307.

Liu, Q., Zuo, S.M., Peng, S., Zhang, H., Peng, Y. *et al.* (2024) Development of Machine Learning Methods for Accurate Prediction of Plant Disease Resistance. *Engineering* 40, 100–110.

Lou, L., Su, J., Ou, X., Li, C., Wang, S. *et al.* (2023) Multi-locus genome-wide association studies reveal genetic loci and candidate genes for leaf angle traits in cut chrysanthemum. *Scientia Horticulturae* 322, 112406.

Lu, H., Giordano, F. and Ning, Z. (2016) Oxford Nanopore MinION sequencing and genome assembly. *Genomics, Proteomics and Bioinformatics* 14, 265–279.

Nicholls, H.L., John, C.R., Watson, D.S., Munroe, P.B., Barnes, M.R. *et al.* (2020) Reaching the end-game for GWAS: machine learning approaches for the prioritization of complex disease loci. *Frontiers in Genetics* 11, 350.

Osborn, T.C., Pires, J.C., Birchler, J.A., Auger, D.L., Chen, Z.J. *et al.* (2003) Understanding mechanisms of novel gene expression in polyploids. *Trends in Genetics* 19, 141–147.

Rehman, A.U., Iso-Touru, T., Junkers, J., Rantanen, M., Karhu, S. *et al.* (2024) Multi-model GWAS reveals key loci for horticultural traits in reconstructed garden strawberry. *Physiologia Plantarum* 176, e14440.

Sattler, M.C., Carvalho, C.R. and Clarindo, W.R. (2016) The polyploidy and its key role in plant breeding. *Planta* 243, 281–296.

Schulz, D.F., Schott, R.T., Voorrips, R.E., Smulders, M.J., Linde, M. *et al.* (2016) Genome-wide association analysis of the anthocyanin and carotenoid contents of rose petals. *Frontiers in Plant Science* 7, 1798.

Slavin, J.L. and Lloyd, B. (2012) Health benefits of fruits and vegetables. *Advances in Nutrition* 3, 506–516.

Soltis, P.S. and Soltis, D.E. (2000) The role of genetic and genomic attributes in the success of polyploids. *Proceedings of the National Academy of Sciences of the United States of America* 97, 7051–7057.

Song, X., Xu, Y., Gao, K., Fan, G., Zhang, F. *et al.* (2020) High-density genetic map construction and identification of loci controlling flower-type traits in *Chrysanthemum (Chrysanthemum × morifolium Ramat.)*. *Horticulture Research* 7, 108.

Sun, S., Dong, B. and Zou, Q. (2021) Revisiting genome-wide association studies from statistical modelling to machine learning. *Briefings in Bioinformatics* 22, bbaa263.

Tam, V., Patel, N., Turcotte, M., Bossé, Y., Paré, G. *et al.* (2019) Benefits and limitations of genome-wide association studies. *Nature Reviews Genetics* 20, 467–484.

Tibbs Cortes, L., Zhang, Z. and Yu, J. (2021) Status and prospects of genome-wide association studies in plants. *The Plant Genome* 14, e20077.

Todesco, M., Bercovich, N., Kim, A., Imerovski, I., Owens, G.L., *et al.* (2022) Genetic basis and dual adaptive role of floral pigmentation in sunflowers. *eLife* 11, e72072.

Uffelmann, E., Huang, Q.Q., Munung, N.S., De Vries, J., Okada, Y. *et al.* (2021) Genome-wide association studies. *Nature Reviews Methods Primers* 1, 59.

Upadhyaya, S.R., Danilevicz, M.F., Dolatabadian, A., Neik, T.X., Zhang, F. *et al.* (2024) Genomics-based plant disease resistance prediction using machine learning. *Plant Pathology* 73, 2298-2309.

Van de Peer, Y., Mizrachi, E. and Marchal, K. (2017) The evolutionary significance of polyploidy. *Nature Reviews Genetics* 18, 411–424.

Venter, J.C., Adams, M.D., Myers, E.W., Li, P.W., Mural, R.J. *et al.* (2001) The sequence of the human genome. *Science* 291, 1304–1351.

Visscher, P.M., Brown, M.A., McCarthy, M.I. and Yang, J. (2012) Five years of GWAS discovery. *The American Journal of Human Genetics* 90, 7–24.

Wamhoff, D., Patzer, L., Schulz, D.F., Debener, T. and Winkelmann, T. (2023) GWAS of adventitious root formation in roses identifies a putative phosphoinositide phosphatase (SAC9) for marker-assisted selection. *PLoS One* 18, e0287452.

Wetterstrand, K.A. (n.d.) DNA sequencing costs: Data from the NHGRI genome sequencing program (GSP). National Human Genome Research Institute. Available at: www.genome.gov/sequencing costsdata (accessed 1 February 2025).

Yang, F., Guo, Y., Li, J., Lu, C., Wei, Y. *et al.* (2023) Genome-wide association analysis identified molecular markers and candidate genes for flower traits in Chinese orchid (*Cymbidium sinense*). *Horticulture Research* 10, uhad206.

Zhang, G.Q., Liu, K.W., Li, Z., Lohaus, R. and Hsiao, Y.Y. *et al.* (2017) The *Apostasia* genome and the evolution of orchids. *Nature* 549, 379–383.

Zheng, T., Li, P., Li, L. and Zhang, Q. (2021) Research advances in and prospects of ornamental plant genomics. *Horticulture Research* 8, 65.

Zhu, Y., Zhu, G., Xu, R., Jiao, Z., Yang, J. *et al.* (2023) A natural promoter variation of *SlBBX31* confers enhanced cold tolerance during tomato domestication. *Plant Biotechnology Journal* 21, 1033–1043.

9 Genetic Transformation of Orchids

Kasi K. Viswanath[1], Jian-Zhi Huang, Pablo Bolaños-Villegas[2,3,4] and Fure-Chyi Chen[1]*

[1]Department of Plant Industry, National Pingtung University of Science and Technology. Pingtung, Taiwan; [2]Fabio Baudrit Agricultural Research Station, University of Costa Rica, Alajuela, Costa Rica; [3]Lankester Botanical Gardens, University of Costa Rica, Cartago, Costa Rica; [4]Faculty of Food and Agricultural Sciences, Rodrigo Facio Campus, School of Agronomy, University of Costa Rica, San Jose, Costa Rica

Abstract

Orchids are prized for their beautiful flowers, but traditional breeding takes time and may not quickly produce new varieties. Genetic transformation can help, though different genera react differently to methods like biobalistics and transfection by *Agrobacterium tumefaciens*. This review examines the pros and cons of these techniques and necessary adjustments, such as adding acetosyringone, chemicals to induce sensitivity to *A. tumefaciens*, selecting vectors, and choosing appropriate explant tissue for transformation. We hope thus to aid research teams across the world in creating innovative traits to meet market demand.

Keywords: Orchids, orchid breeding, orchid flower color, orchid stress tolerance, genetic transformation.

9.1 Introduction

Orchids, the most elegant of all ornamentals, hold a special place in the horticultural and floricultural industries all over the world. According to da Silva (2013), orchid blossoms are highly sought-after due to their remarkable diversity in shape and color, fragrant flowers, and the long lifespan of flowers, that in some cases extends for several months. Almost one third of monocotyledons and approximately 10% of blooming families are orchids, thus this is the largest family of angiosperms with over 880 genera and 30,000 species (Dressler, 2005; Govaerts *et al.*, 2021). Moreover, according to Leitch *et al.* (2009), genomes that belong to the orchid family are typically large and vary by as much as 168 times (1C = 0.33–55.4 pg, where the C value refers to the DNA content found in a haploid genome), indicating a remarkable evolutionary diversification that has allowed them to occupy every ecological niche on Earth (Vitt *et al.*, 2023).

Many orchid blossoms, both wild and hybrids, have been utilized in cuisine, perfumes, gardening, and traditional medicine (Gutierrez, 2010; Chinsamy, 2014; Bhattacharyya *et al.*, 2014). Orchids are sold as either cut flowers or potted plants, and their combined output now accounts for about 8% of the global floriculture market (Chugh *et al.*, 2009), moreover the cultivation of these valuable plants generates precious foreign exchange for many developing countries (Laws, 1995; Tomokuhara and Mii,

*Corresponding author: fchen@mail.npust.edu.tw

 Biotechnology of Orchids and Other Ornamentals (Eds P. Bolaños-Villegas and F-C. Chen)
DOI: 10.1079/9781800627895.0009

2001). As recently as 2023 the total market value of orchids was estimated at US$751.9 million, a sizeable amount (Zion Market Research, 2024).

Plastid phylogenomics suggest that the Orchidaceae is composed of five subfamilies: Apostasioideae, Vanilloideae, Cypripedioideae, Orchidoideae (composed of Orchidae, Codonorchidae, Cranichidae, and Diuridae) and Epidendroideae (composed of Sobralieae, Mallaxideae, Epidendreae, Vandeae, and Cymbidieae, among others). Apostasioideae, Vanilloideae and Cypripedioideae (including *Phragmipedium*) have the fewest genera and species among these five subfamilies (Serna-Sánchez *et al.*, 2021). Figure 9.1 depicts the floral diagrams of several orchids, such as *Phalaenopsis* (the moth orchid), *Oncidium* (the dancing lady orchid), and *Cymbidium* (the boat orchids).

The remarkable beauty, exquisite look, and long lifespan of orchid hybrids have contributed to their strong appeal among producers and consumers. One of the main aims in orchid cross pollination and hybridization is to create new commercial cultivars with enhanced anatomical features, the ability to withstand environmental and shipping stress, and to better resist disease (Azadi *et al.*, 2016). However, due to their lengthy juvenile phase and long reproductive cycle, orchids are often not amenable to quick selection of showy new traits through traditional breeding methods (Yang *et al.*, 1999; Chai and Yu, 2002).

Next-generation sequencing (NGS) and bioinformatics are technologies that have made tremendous strides and have produced a wealth of sequence data (Lee, 2023). And because of genomics and the transcriptome study of the orchid genome the first orchid databases were created (Fu *et al.*, 2011; Su *et al.*, 2013;). Moreover, the technology for the genetic transformation of orchids is a useful approach for the study of gene function, which has also been shown to be a safe and effective alternative to breeding new orchid varieties. Thus, by combining genomics with transformation it may be possible to move forward with studies regarding defense mechanisms and floral development at the molecular

Fig. 9.1. Floral morphology of orchid genera where genetic transformation has been implemented. A) *Phalaenopsis* DO Black Beauty 'F4528'. B) *Cattlianthe* 'Lo's White World'. C) *Odontocidium* Catante 'Pacific Sun Spot'. D) *Vanda* 'Donchai x Happiness'. E) *Cymbidium* Mary Green 'Piccadilly Circus'. F) *Dendrobium* D05 UCR G) Simplified diagram showing orchid floral organs in *Phalaenopsis*. Images taken at the Lankester Orchid Gardens of the University of Costa Rica in Cartago province except for figures A and B, taken at the Hoya nursery in Pingtung County, Taiwan, and F, taken at the Fabio Baudrit Agricultural Station of University of Costa Rica in Alajuela Province. The orchid in figure G was designed with Adobe Express.

level (Yu *et al.*, 2001; O'Neill, 1997; Zhang and O'Neill, 1993; Huang *et al.*, 2015, 2016; Wang *et al.*, 2019).

This report seeks to present an extensive overview of orchid transformation techniques documented to date, including examples from other plant species to facilitate future functional genomics research. Each technique is detailed in clear and straightforward language to promote swift adoption by laboratories around Asia and the world.

9.2 Types of Transformation Techniques

Genetic transformation can introduce new traits in orchids through various techniques. Three primary methods are considered effective for the transformation process:

1. *Agrobacterium tumefaciens*-mediated transformation.
2. Biolistics or microprojectile bombardment.
3. Direct gene transformation.

To date only two transformation methods, microprojectile bombardment and *Agrobacterium tumefaciens*-mediated transformation, have been applied to orchids with satisfactory results (Nan and Kuehnle, 1995; Chia *et al.*, 2001; Sanjaya and Chan, 2007; Shrestha *et al.*, 2007; Mii and Chin, 2010; Hsing *et al.*, 2016). And when it comes to stable gene transfer, *Agrobacterium tumefaciens*-mediated transformation is superior to microprojectile bombardment (Fang *et al.*, 2002; Shou *et al.*, 2004).

9.2.1 Transformation efficiency in orchids

Conventional hybridization breeding approaches may have been the method of choice for many years in the selection of attractive and profitable orchid genotypes. Nonetheless gene transformation is comparatively quick. Two elements are essential for orchid transformation. Selecting the right species and selecting the right tissue for a high capacity for regeneration *in vitro* (Liao *et al.*, 2011).

For instance, unlike other higher plants, mature orchid embryos lack the primordia for leaves and roots (Arditti, 1992; Yeung, 2022). Nonetheless because protocorm-like bodies (PLBs) have a high ability for regeneration compared to regular protocorms, the induction of PLBs and the occasionally protocorm is a crucial strategy for orchid gene transformation (Arditti and Ernst, 1993; Cardoso *et al.*, 2020; Tiwari *et al.*, 2024).

Second, another aspect that may reduce the overall transformation efficiency might be polyploidy (Hsing *et al.*, 2016; Bolanos-Villegas and Chen, 2022), and polyploidy may account for the small number of orchids that have undergone stable gene transformation. Nonetheless, the particle bombardment approach may still yield the greatest number of transformants followed by *A. tumefaciens*-mediated transformation.

9.2.2 Plasmid/vectors, selectable markers

The transformation mechanism and the corresponding plasmids utilized in orchids are displayed in Table 9.1. The efficient selection of transformed explants relies on the use of selectable markers or reporter genes during the selection process. The markers green fluorescent protein (*GFP*) and β-glucuronidase (*GUS*) are cited as the reporter genes of choice in most reports, whether during *A. tumefaciens*-mediated transformation or by particle bombardment (Chia *et al.* 1994; de Silva *et al.* 2013; Kui *et al.* 2017), although the firefly luciferase gene (*luc*) has also been used successfully as a reporter gene (Chia *et al.* 1994).

Antibiotic resistance genes that encode resistance proteins, such as the hygromycin phosphotransferase II gene, also known as *hptII*, and the neomycin phosphotransferase II gene, also known as *nptII*, are also used as selection markers for orchids during the transformation process. These genes mediate the degradation of kanamycin and hygromycin, two common antibiotics used during explant selection (Sundar and Sakthivel, 2008). Experimental results suggest that in monocot cells and tissues sensitivity to kanamycin is low, thus due to its low concentration in the medium and short selection time, hygromycin has been the preferred antibiotic for the transformation of orchids (Yu *et al.*, 1999).

Table 9.1. Genetic transformation methods for different orchids, and relevant adjustments. Author's own table.

Plant	Explant source	Vector used	Mode of transformation	Gene of interest/ reporter	Reference
Brassia, Cattleya and *Doritaenopsis*	PLBs	pDPG165	Particle bombardment	β-glucuronidase, luc (luciferase)	Knapp *et al.*, 2000
Cymbidium	PLBs	pKH200	Particle bombardment	β-glucuronidase	Yang *et al.*, 1999
Dendrobium phalaenopsis and *D. nobile*	PLBs, calli	pCAMBIA1301	Particle bombardment	β-glucuronidase	Men *et al.*, 2003a
Dendrobium	PLB, thin sections	pBI121-DOH1	*A.tumefaciens* LBA4404	β-glucuronidase	Yu *et al.*, 2001
D. nobile	PLBs	pCAMBIA1301	*A.tumefaciens* EHA105, AGL1	β-glucuronidase	Men *et al.*, 2003b
Oncidium	PLBs	pCAMBIA1301	*A.tumefaciens* LBA4404 and EHA105	β-glucuronidase	Liau *et al.*, 2003a
Oncidium	PLBs	pSPFLP	*A.tumefaciens* EHA105	*Pflp* (ferredoxin-like protein)	Liau *et al.*, 2003b
Phalaenopsis	PLBs	pCAMBIA1304	Particle bombardment	Green fluorescent protein, β-glucuronidase	Chan *et al.*, 2003
Phalaenopsis	Cell clumps	pIG121Hm, pTOK233	*A.tumefaciens* LBA4404	β-glucuronidase	Belarmino and Mii, 2000
Phalaenopsis	PLBs	pCAMBIA1301	*A.tumefaciens* EHA105	β-glucuronidase	Chan *et al.*, 2003
Phalaenopsis	PLBs	pTOK233	*A.tumefaciens* LBA4404	β-glucuronidase	Chai *et al.*, 2002
Phalaenopsis	PLBs	pB1221-GUS	Particle bombardment	β-glucuronidase	Anzai *et al.*, 1996

Knapp *et al.* (2000) employed the phosphinothricin acetyltransferase gene, also known as *bar*, as a selection marker, and Bialaphos (phosphinothricyl-alanylalanine, $C_{11}H_{22}N_3O_6P$), an alanine derivative of phosphinothricin (PPT), or even L-methionine- S-sulfoximine ($C_5H_{12}N_2O_3S$, also known as MSO) (Bolton *et al.*, 2011), to select for this marker. An additional benefit of the transformation *bar* gene in orchids is that it can function as a selection gene for resistance to glufosinate ($C_5H_{12}NO_4P$) as well (Anzai and Tanaka, 2001; Bolton *et al.*, 2011).

In contrast to selection markers for antibiotic resistance gene, the sweet pepper ferredoxin-like protein (*pflp*) has been shown to be an efficient selection marker for the transformation of *Oncidium* orchids (Liau *et al.*, 2003a). One key advantage of using this marker is that it only takes two weeks to identify potential transformants (Teixeira da Silva *et al.*, 2011). Without the use of any antibiotics or herbicides, Thiruvengadam *et al.* (2011) reported an alternative positive selection technique for the selection of orchid transformants. Using this method, transgenic orchids are created by delivering the phosphomannose-isomerase gene (*pmi*), which only allows transformed cells to proliferate on a chosen agent.

9.2.3 Response of co-cultivation methods on gene transformation efficiency

Depending on the method, co-cultivation may take varying amounts of time. Until about 20 years ago, it was believed that monocots could not be transformed by *A. tumefaciens* directly. This belief may have been sparked by the fact that these plants are not within the host range of this bacterium and that monocots lack the ability to divide their cells effectively at wound sites (Hiei *et al.*, 2014). Moreover, it is the accumulation of phenolic compounds released from dicot plant tissues that activate the virulence (*vir*) genes in *A. tumefaciens* that allow it to attack and transfect wound sites (Stachel *et al.*, 1985).

To compensate for the absence of phenolic compounds in monocots, it is advisable to add exogenous phenolics such as acetosyringone (AS), to the bacterial and/or co-cultivation media to trigger the expression of the *vir* genes that are responsible for T-DNA transfer (Stachel *et al.*, 1985). Nonetheless since orchids are monocot species, they typically resist infection by *A. tumefaciens*. To overcome this initial resistance the expression of *vir* genes must be enhanced by having AS included in the co-cultivation media or integrated into the bacterial medium well before co-cultivation. Results suggest that at least in *Dendrobium* orchids, additional inducers of *vir* expression are coumaryl alcohol and the phenolic compound coniferyl alcohol (Nan *et al.*, 1997). Also, according to Subramaniam and Xavier (2010), temperature, pH, L-cysteine, and silver nitrate have all been shown to have an impact on orchid transfection.

The standard procedure for *A. tumefaciens* transformation with the required gene construct involves growing the organism on either *Agrobacterium* minimal medium (AB) or yeast extract beef (YEB) medium with the desired antibiotics, overnight, and at 28°C. After dipping the calli/PLB in the *A. tumefaciens* suspension for 10 to 30 minutes, it has to be air-dried and plated onto the co-cultivation medium, where it has to be kept in the dark for 2 to 3 days. The tissue is then placed on antibiotic-containing screening medium to kill excess bacteria and to initiate the selection of transformants. This culture step may take 1 or 2 weeks to apply selection pressure while allowing the growth of those cells that may have undergone transformation.

9.3 Options to Enhance *A. tumefaciens*-mediated Transformation

When *Agrobacterium tumefaciens* is used for genetic transformation, the integration of single transgenic sequences into the plant genome is verifiable and its expression is usually stable. However, the early years of transforming monocots using *A. tumefaciens* were sluggish and plagued with erratic results (De Cleene and De Ley, 1976). Since then, several methods have been validated and have allowed researchers to increase the relative transformation efficiency in several species. Occasionally, the single most key factor is the choice of the right *A. tumefaciens* strain. For instance, Ishida *et al.* (2007) created a binary vector for maize, which was effectively delivered using the *A. tumefaciens* strain LBA4404. Other top strains of *A. tumefaciens* for genetic transformation are AGL1 in barley, EHA101 and EHA105 in wheat, and LBA4404 and EHA101 in rice (Hiei *et al.*, 2014).

Following the choice of an effective strain it is also advisable to induce hypersensitivity to *Agrobacterium*. For instance, inhibitors of purine synthesis, including 5-Fluorouracil ($C_4H_3FN_2O_2$, also known as 5-FU and Fluoroplex), Mizoribine ($C_9H_{13}N_3O_6$ also known as Bredinin), Azaserine ($C_5H_7N_3O_4$, also known as L-Serine) and Avicin G ($C_{98}H_{155}NO_{45}$) (Bolton *et al.*, 2011), were successfully employed by Roberts *et al.* (2003) for 16 hours to induce hypersensitivity to *A. tumefaciens* in tobacco, *Arabidopsis*, and *Ageratum* (also known as whiteweed). Several plants in that report showed a 180-fold increase in transformation efficiency by boosting hypersensitivity. It was postulated at the time that the use of either Avicin G or Azaserine disrupts both the purine and pyrimidine pathways, and may lead to the highest effect (Roberts *et al.*, 2003). Another interesting case is the Easter lily (*Lilium longiflorum*), which was believed to be resistant to transformation mediated by *A. tumefaciens*, and up to a point was only amenable to the particle bombardment approach (Lipsky, A. *et al.*, 2000). Nonetheless, *A. tumefaciens* was eventually shown to mediate transformation in lilies by Hoshi *et al.* (2004). This feat was accomplished through the removal of key macronutrient ammonium nitrate, from the culture media, and through thorough scratching of calli prior to co-cultivation. This is done by using sandpaper to injure the calli prior to infection (Hoshi *et al.*, 2004). This team also discovered that the presence of ammonium nitrate in the co-cultivation medium inhibits the temporary expression of the *GUS* gene in lily (Hoshi *et al.*, 2004). The presence of other nutrients may have positive effects too. For instance, Wu *et al.* (2014) demonstrated that including copper sulfate and 6-benzylaminopurine into the transfection media significantly improves the transformation rate in *Sorghum*.

Additional adjustments are possible. For instance, according to Montoro *et al.* (2003), using calcium-free media significantly increases the number of transformation events in plant and bacterial tissue culture preparations of rubber trees (*Hevea brasiliensis*). Also, according to Olhoft *et al.* (2001), adding L-cysteine and thiol compounds to the co-cultivation media may enhance *Agrobacterium* transfer DNA Ti Plasmid (T-DNA) delivery into soybean cotyledon node cells. By inhibiting wound and plant pathogenic-induced responses, L-cysteine and other thiol compounds might increase the susceptibility of cotyledon node cells to infection by *A. tumefaciens* and, consequently, raise the potential for transformation in these totipotent soybean cells. According to research by Zambre *et al.* (2003), light also significantly accelerates the transfer of *A. tumefaciens* genes into plant cells. This is significant since in most projects that involve transformation, co-cultivation is typically carried out for two to three days in the dark. Nonetheless, Zambre *et al.* (2003) described the incubation of two plant species—*Phaseolus* and *Arabidopsis*—in either a 16/8 photoperiod or a continuous light regime throughout the co-cultivation period. Remarkably they reported a significant increase in T-DNA transfer. Thus, by using any of the aforementioned techniques either alone or in combination, it may be possible to boost the transformation rate in orchids. In our case we have observed encouraging results regarding these techniques.

Someya *et al.* (2013) described how to use *A. tumefaciens* to express the coding sequence for the 1-aminocyclopropane-1-carboxylic acid (ACC) deaminase, which when expressed enhanced the process of transferring T-DNA into *Solanum lycopersicum* (tomato) and *Erianthus ravennae* (the hardy Pampas grass). According to the authors, plant cells may produce less ethylene when the ACC deaminase is expressed, which may improve T-DNA delivery into the nucleus. This may be a wise approach to orchid transformation that needs careful consideration.

The cultivar of choice may influence the final transfection rate as well. For instance, in rice, Saika and Toki (2010) found that the *indica* rice cultivar Kasalath had a tenfold higher transformation efficiency compared to *Nipponbare*. Also, Hisano *et al.* (2017) reported that in barley, the production of transformation-efficient lines can be facilitated by the introgression of several transformation amenability alleles from the 'Golden Promise' cultivar. Often *in vitro* transfection is not effective in perennial ryegrass (*Lolium perenne*). To address this problem, Patel *et al.* (2013) employed high maltose and heat shock treatments sequentially. Their results suggest that co-cultivation of *A. tumefaciens*-infected calli on a high maltose medium (6%) and then warming them up for 3 minutes at 42°C might boost the transfection efficiency. Gurel *et al.* (2009) reported a similar finding regarding a heat shock treatment during the transfection of *Sorghum*. The authors suggest that the ideal heat treatment for immature embryos is 43°C for 3 minutes before the initiation of transfection by *A. tumefaciens*, which increases the transformation rate. In citrus, one way to significantly boost transfection (15-fold) is to express the maize *knotted1* (*kn1*) gene with the help of the CaMV 35S promoter (15-fold) (Hu *et al.*, 2016).

One key factor that mediates the interaction between *A. tumefaciens* and plant tissues is how the plant host reacts to injury (Potrykus, 1991). Many different wounding techniques have been adopted, including micro-bombardment (Bidney *et al.*, 1992; Knittel *et al.*, 1994; Brasileiro *et al.*, 1996), shaking with glass beads (Grayburn and Vick, 1995), scraping with carborundum (Cheng *et al.*, 1996), or using a microsyringe prior to bacterial inoculation. We believe all these techniques may show a high transformation rate.

9.4 Transformation in Specific Orchid Genera

Improving transformation efficiency is crucial for the development of new orchid varieties. Unfortunately, researchers have made limited advancements in the orchid gene transformation method over the last 20 years. We believe that this is a result of not choosing the right tissues. In Taiwan, the best results have been obtained with PLBs due to their unmatched high capacity for regeneration, especially in *Oncidium*, *Dendrobium*, *Cymbidium*, and *Phalaenopsis* (Hsiao *et al.*, 2011; Huang *et al.*, 2014).

9.4.1 *Oncidium*

To date few studies on *Oncidium* transformation using *A. tumefaciens*-mediated techniques have been published (Liau *et al.*, 2003a,b; You *et al.*, 2003). Liau *et al.* (2003) from the Academia Sinica in Taipei pioneered the active transformation approach for gene transfer into PLBs, significantly contributing to new *Oncidium* cultivars. This team pioneered orchid transformation in many ways. First, the vector construct they used included the coding sequence for *GUS* reporter and showed concluding evidence of its stable expression (Liau *et al.*, 2003a). Second, they successfully transfected PLBs of *Oncidium* by using the *A. tumefaciens* strain EHA105 to insert several reporter genes such as *gusA* (β-glucuronidase), *hph* (hygromycin phosphotransferase), and the cDNA sequence for the *Capsicum annuum* L. ferredoxin-like protein (*pflp*), which had been reported to induce antimicrobial activity against *Pseudomonas syringae* by delaying the hypersensitive response in non-host plants (Liau *et al.*, 2003b). The resulting transgenic *Oncidium* showed resistance to the gram-negative pathogen *E. carotovora*, which causes the soft rot disease (Liau *et al.*, 2003b). By using the *pflp* gene as a selection marker and *E. carotovora* as a selection agent for *A. tumefaciens*, plus delivery by particle bombardment, You *et al.* (2003) developed an effective method for the selection of transgenic orchids that included infection and then co-culture of *Agrobacterium* and orchid PLBs along with tobacco culture cells that release phenolics onto the medium. The resulting plants showed resistance to a pathogenic strain of *E. carotovora* previously isolated by the Asian Vegetable Research and Development Center (AVRDC) in Tainan County. Then Li *et al.* (2005) from Cheng Kung University developed an even more effective procedure by particle bombardment and a clever sucrose osmotic pretreatment that stimulates single-cell embryogenesis (and thus gives rise to non-chimeric, homogeneous plantlets) to create stable *Oncidium* Sharry Baby AM/AOS 'OM8' transgenic plants.

A significant issue in orchid breeding is early flower wilting caused by ethylene. In order to induce ethylene insensitivity in *Oncidium* flowers, Raffeiner *et al.* (2009) used an *A. tumefaciens*-mediated technique (strain EHA105) to transfer an allele of the ethylene receptor gene (*etr1-1*) from *Arabidopsis* to PLBs of *Oncidium* 'Sweet Sugar' at a rate of 2,7%. Then Thiruvengadam *et al.* (2011) at Chung Hsing University in Taichung developed another transfection method using the phosphomannose-isomerase (*pmi*) gene for selection of transgenic plants on mannose selective medium for *Oncidium* Gower Ramsey. The transformation rate was boosted dramatically to 21%, while the regeneration time of plantlets was cut by 50% (8 to 4 months) although no specific new traits were introduced at the time.

Other traits of interest in *Oncidium* are early flowering and more flowers per plant. Thiruvengadam *et al.* (2012) achieved this in *Oncidium* Gower Ramsey by co-cultivating PLBs with *A. tumefaciens* strain LBA4404. Transfection allowed delivery of a pBI121 binary vector that carried a copy of the *Oncidium* equivalent of the Arabidopsis *AGAMOUS* gene (*OMADS1*), which regulates stamen and carpel development, driven by the 35S promoter. Selection was performed on 45-day-old PLBs on plates containing 200 mg l^{-1} kanamycin, for three weeks. The transformation rate was 9% and the resulting plants quickly flowered after 24 months, instead of 36 months in the wild type, formed 12 pseudobulbs per plant instead of 5 in the wild type, and developed 50 flowers per inflorescence versus only 32 in the untransformed wild type (Thiruvengadam *et al.*, 2012). Taken together, while the precise function of the *Oncidium MADS1* gene was not ascertained, the corresponding overexpression phenotype was of commercial significance.

Another interesting phenotype was obtained in *Oncidium* when *A. tumefaciens* allowed the delivery of RNAi construct for the phytoene synthase (*PSY*) for carotenoid synthesis into PLBs (Liu *et al.*, 2014). In this report, the authors found that the geranyl synthase gene and *PSY* were both down-regulated, and the resulting transgenic *Oncidium* hybrid plants grew vivid green leaves and showed a semi-dwarf phenotype of commercial value.

9.4.2 *Phalaenopsis*

One of the most aesthetically pleasing and profitable orchid crops is *Phalaenopsis*, the moth orchid. Quite a body of work exists regarding its transformation. For instance, Belarmino and Mii

(2000) reported using *A. tumefaciens* to transform tissues in *Phalaenopsis* with the sequence for the *GUS* reporter gene. In that report, co-cultivation was done in two steps. First, a suspension of *A. tumefaciens* was incubated in New Dogashima medium (NDM) (Tokuhara and Mii, 1993) for 10 hours, at 30 rpm agitation and at 28°C. Following a wash in sucrose-free NDM medium, the cells were plated on solid medium. AS at a concentration of 500 μM was also included in the solid medium, and co-cultivation proceeded for three days in the dark. Results suggested that incubation for 10 hours produced the highest expression of *GUS* at rate of 10% per 100 mg of cell clumps. On co-culture medium, different concentrations of AS were also tested, and 500 μM was discovered appropriate for bluer spots/100 mg of cell clumps. In order to determine the efficacy of *Phalaenopsis* transformation, Chan *et al.* (2003) attempted three distinct methods: particle bombardment, *A. tumefaciens*, and sonication-assisted *A. tumefaciens* method (SAAT). Based on previous research by Hsieh *et al.* (1995), a filter paper base above the medium was added to improve the transformation during bombardment; however, no discernible improvement was seen after bombardment compared to the absence of the filter paper foundation. Another approach involved injuring the PLB and then infiltrating it with *A. tumefaciens* by vacuum, however this did not result in a change in *GUS* expression. Additionally, they performed SAAT, which is co-cultivation of the explants after soaking them in *A. tumefaciens* and sonicating them for different lengths of time. In comparison to the first two approaches, a 150-second sonication followed by co-cultivation improved the transient expression of *GUS*. Therefore, they proposed that temporary *GUS* expression with SAAT was more effective for transformation in *Phalaenopsis* than either a particle gun or *A. tumefaciens*. A similar approach was adopted by Mishiba *et al.* (2005) who successfully transformed immature protocorms of *Phalaenopsis* obtained after 21 days of culture on liquid NDM. The protocorms were inoculated with *Agrobacterium* harboring both the *GUS* reporter and hygromycin selection markers, and maximum transformation efficiencies (1.3–1.9%) were observed. Then Semiarti *et al.* (2007) in Japan created a practical transfection method with *A. tumefaciens* LBA4404 for protocorms of the wild orchid species *P. amabilis*. Protocorms were transformed with the binary vector pG-35SKNAT1 in which expression of shoot apical meristem (SAM) regulator BP/KNAT1 is driven by the CaMV 35S promoter, along with the terminator for the nopaline synthase. The transformation rate was 0.3, and was estimated from tolerance to kanamycin, PCR and the mutant phenotype conferred by the *BP/KNAT1* sequence, in which protocorms grow many abnormal shoots and aberrant leaves. In Indonesia and Japan, Subramaniam *et al.* (2010) developed an effective transformation strategy for *Phalaenopsis violaceae*. In this report it was verified that strain EHA105 performs better than EHA101, as suggested by high expression of the *GFP* reporter gene in PLBs. Also, stable transformants could be identified by co-cultivating those PLBs in the dark, at 24°C, in a half-strength Murashige & Skoog (MS) medium supplemented with 200 mg l^{-1} L-cysteine, 60 μM silver nitrate, and 5% of the banana pulp. Then Semiarti *et al.* (2014) in Indonesia, created a straightforward and safe method for transferring T-DNA into *P. amabilis* and *Spathoglottis plicata*. In this method pollinia (e.g., pollen masses) are first immersed into bacterial culture for 30 minutes, then dried up, and then used to pollinate orchid flowers. The resulting seeds were grown onto 10 mg l^{-1} hygromicin. The presence of the transgene in PLBs was by PCR combined with a histochemical *GUS* assay. Transformation rates were 2.1 % in *P. amabilis*, and 0.53% in *S. plicata*.

For *Doritaenopsis* (now *Phalaenopsis*) a similar system of transfection of pollinia was devised by Tsay *et al.* in 2012, at Chaoyang University. The system consisted of a pollen-mediated activation tagging system for the random creation of T-DNA mutants that could be selected on hygromycin. The report indicates that transfection was effective at a low concentration of *A. tumefaciens* strain EHA105 (0.5–1 × 10^8 CFU ml^{-1}), at 0.1 mM of AS, and only 6 hours of co-culture. A total of 30 transformants of *P. amabilis*, 7 transformants of *Dtps*. Sinica Sunday, and 15 transformants of *Dtps*. Taisuco Firebird were obtained after selection in 5 mg l^{-1} hygromycin. Results were reconfirmed by PCR (Tsay *et al.*, 2012).

Hsing *et al.* (2016) at Academia Sinica in Taipei devised an effective transformation strategy for *Phalaenopsis* that involved transfection of

PLBs, artificial pollination and backcross. In this report the authors chose several varieties for transfection including the tetraploid *Dtps.* Join Angel 'TH274-1', the tetraploid *P. aphrodite* 'M1663' and the diploid *P. aphrodite* subsp. *formosana.* And for backcrossing, the lines chosen were the two *P. aphrodite* varieties, plus hybrids C2-4 (TH274 x *Phalaenopsis equestris*), and V65 x V15 (by Tailing Co., Tainan). The vector for transfection included the sequence for the green fluorescence marker gene *eGFP* driven by the ubiquitin promoter, plus the selection marker *HptII* for hygromycin resistance driven by the CaMV 35S promoter. The *Agrobacterium* strain used was EHA105. Only the tetraploid *P. aphrodite* 'M1663' was transformed and backcrossed. The hygromycin resistant explants varied from 43 to 56% in four the first backcross generation (BC_1) progenies, however expression of *eGFP* reached only 5.2%, which is still twice the regular rate.

9.4.3 *Dendrobium*

As one of the earliest orchid genera to undergo transformation, *Dendrobium* holds a special place in orchid history. It was employed to evaluate *Agrobacterium*-mediated transformation, achieving partial success as only a fragmentary T-DNA sequence was incorporated (Nan *et al.*, 1998). Then in 2015 the complete genome sequence of *Dendrobium officinale* Kimura et Migom was published (Yan *et al.*, 2015), opening the door for the use of *Dendrobium* as a research model for orchid genetics (Yan *et al.*, 2015).

Back in 2001 in Singapore, Yu *et al.* (2001) transfected sectioned PLBs of *Dendrobium* Madame Thong-In (*Dendrobium* Somsak x *D.* Suzie Wong) with *Agrobacterium tumefaciens* strain LBA4404. This strain carried a pBI121 binary vector that featured an antisense sequence for the putative *DOH1* homeobox gene, the sequence for the *GUS* β-glucuronidase, and a sequence for the neomycin phosphotransferase, all driven by the CaMV 35S promoter. Selection occurred after 8 months of cultivation on medium supplemented with 200 mg l^{-1} of kanamycin and carbenicillin. Out of 190 orchid plantlets on kanamycin medium, 154 (81%) were suspected to be transformants based on defects in plant architecture, and this later was confirmed by PCR.

Then in 2003, in Shenzhen, China, the team of Suzhen Men (Men *et al.*, 2003a) reported a protocol for the transfection of *Dendrobium nobile* PLBs using *Agrobacterium* strains AGL1 and EHA105, with both carrying the binary vector pCAMBIA1301. This vector features the sequences for the *GUS* reporter gene, and the *hpt* gene (as selection markers) driven by the CaMV 35S promoter. Sectioned PLBs were infiltrated for 30 min and then co-cultured for three months with *A. tumefaciens* in half-strength MS medium containing 100 μM AS, 30 mg l^{-1} of hygromycin and 250 mg l^{-1} of cefotaxime (only for 2 months).

The authors reported that of 578 sliced PLBs, 73 transformed explants were verified, for a transformation rate of 18%.

A similar result was reported in Malaysia by Subramaniam *et al.* (2009) for PLBs of *Dendrobium* 'Savin White'. Three-month-old PLBs from shoot tip culture were first grown on full-strength MS medium, and then sub-cultured for one month on half-strength MS medium supplemented with vitamins and 1 mg l^{-1} of 6-benzylaminopurine (BAP). Then these PLBs were transfected with strain EHA105 and co-cultured in the presence of the AS. A concentration of 150 μM of AS leads to the highest expression of the *GUS* reporter, while co-cultivation for 3 days is also beneficial. Other parameters that positively influence transfection were calcium concentration in the MS medium, concentration of silver nitrate ($AgNO_3$), immersion period, and bacterial density (OD600) (Subramaniam *et al.*, 2009).

Breeding *Dendrobium* varieties with new colors has been a top priority in spring *Dendrobium*, which grows large and thick white flowers (Quan *et al.*, 2011). New colors might be integrated into *Dendrobium* with the help of a transgene for the chalcone synthase (*CHS*), a gene that regulates the synthesis of flavonoids and anthocyanins and thus imparts showy coloration patterns (Quan *et al.*, 2011). In this case Quan *et al.* (2011) at the Shanghai Institute of Technology tested the SAAT method of Chan *et al.* (2003) for the delivery of *CHS* into PLBs of spring *Dendrobium* 'Sanya'. But this time these Sanya PLBs were thoroughly sonicated for 10 min at 40 KHz (35 W) and then were immersed in *Agrobacterium* suspension for 60 min. The *Agrobacterium* strain LBA4404 was engineered to carry the binary vector AG205 with the sequence for *CHS* cloned from *Petunia hybrida*, plus the *nptII* gene as marker for kanamycin resistance. The transformation rate was 0.5% and was confirmed by PCR, RT-PCR and Southern Blot (Quan *et al.*, 2011).

Another successful case was that of *Dendrobium* Chao Praya Smile, in which the coding sequence for the MADS-Box factor *OVEREXPRESSION OF CONSTANS 1* (*SOC1*)-like gene, *DOSOC1*, was successfully overexpressed by Ding *et al.* (2013) at the National University of Singapore. In that report *Agrobacterium* successfully delivered the binary vector pGreen 029-35S, in which the expression *DOSOC1* is driven by two CaMV 35S promoters. These transgenic *Dendrobium* lines consistently displayed early flowering, a useful trait.

In Malaysia, the team of Uddain, *et al.* (2015) used mildly wounded PLBs of *Dendrobium* Broga Giant (*Dendrobium* Bobby Messina x *D. superbiens*), to quantify the overall efficiency of *Agrobacterium*-mediated transformation with strain LBA4404, bacterial chemotaxis (e.g., attraction to plant host tissues) and temporary *GUS* expression. Although transformation was confirmed no obvious phenotypes were characterized or developed.

Highly efficient transfection of *Dendrobium nobile* (27%) was reported by Phlaetita *et al.* (2015) at Chiba University in Japan after co-culture of 21 day-old protocorms for 3 days with the *Agrobacterium* strain EHA101. This strain carries the pIG121Hm binary vector, which contains a hygromycin phosphotrasferase gene (*hpt*) and an intron *GUS* gene, both under the control of the CaMV 35S promoter, plus *nptII* under the control of the nopaline synthase promoter (Phlaetita *et al.*, 2015). In this report, transfected PLBs were grown on ND medium containing hygromycin (30 mg l^{-1}) and meropenem (10 mg l^{-1}). After secondary selection, PLBs were transferred onto ND medium containing hygromycin (10 mg l^{-1}) and meropenem (10 mg l^{-1}) for shoot formation and rooting. Putative transgenic plants were transferred to pots and grown in a growth chamber at 25°C under a 16 hour photoperiod without any acclimatization treatment, and fortunately many plantlets were successfully regenerated. No obvious phenotypes were recorded though.

Then in 2017 at the National Orchid Conservation Center in Shenzhen, Kui *et al.* did transfection on the medicinal plant *Dendrobium officinale* to test the pCambia-1301-35SN binary vector which features the 35S promoter plus *GUS* and the Superfolder green fluorescence protein (SG) as reporter genes. It was possible to validate the efficiency of several new promoters such as the mirabilis mosaic virus promoter (*MMV*), the cassava vein mosaic virus promoter (*CVMV*), and the peanut chlorotic streak caulimovirus promoter (*PCISV*). Moreover, genomic editing was attempted on several target genes in the lignocellulose biosynthesis pathway, and in several cases it was a success.

Finally, Sawettalake *et al.* (2017) resorted to *Agrobacterium*-mediated transformation on *Dendrobium* Chao Praya and effectively transferred an *APETALA1* (*AP1*) ortholog called *DOAP1*. The result was *35S:DOAP1* transgenic orchids that flowered earlier but developed fewer flowers than the wild type.

9.4.4 *Cymbidium*

Genetic transformation of *Cymbidium* orchids has not been pursued to the same extent as in *Phalaenopsis*. Some of the earliest work was carried by Yang *et al.* (1999) at Chungbuk National University in Korea and consisted of testing bombardment to develop transgenic *Cymbidium* orchid lines. In that report the vector of choice was pKH200, which features *GUS* and *nptII* for selection. Then Chin *et al.* (2007) at Chiba University used sliced PLBs (0.5 mm) of three genotypes (lines RY, L4 and L23) kept in a liquid ND medium (10 or 30 g l^{-1} sucrose plus 10 mg l^{-1} meropenem) to test transfection with strain EHA101 and vector pIG121Hm. It was determined that the addition of AS (100 μM) during cocultivation, plus the genotype (L4) have a profound impact on the transformation rate.

A transformation system for PLB slices (3 mm) of *Cymbidium hybridum* with *Agrobacterium* strains EHA105 and LBA4404 was established by Wang *et al.* (2011) at Tongji University in Shanghai. The binary plasmid vector pCAMBIA1301 was used. It was determined that for LBA4404 the best density (OD600) was 0.6, and that selection with hygromycin gave the best results at a concentration of 5 mg l^{-1}.

The transformation rate was 62%, and 13% of those explants may have survived. Another suitable method for *Agrobacterium*-mediated transformation of *Cymbidium sinensis* was also developed by Xie *et al.* (2015) using rhizomes grown from the culture of shoot-tips from *Cymbidium sinensis* cv. Qijianbaimo. The highest expression of the *GUS* marker was observed (11%) when rhizomes pre-cultured for 39 days were soaked in bacterium suspension supplemented with 200 μmol l^{-1} AS for 35 min, followed by cultivation on medium supplemented with the same concentration of AS for 7 days. The transformation rate was reported at 0.75%.

9.4.5 *Brassia* and *Cattleya*

To our knowledge, to date there are not any reports of transformation in *Brassia* orchids by *Agrobacterium*. Nonetheless *Brassia*, *Cattleya*, and *Doritaenopsis* were the first three orchid genera that Knapp *et al.* (2000) transformed by bombardment using the *bar* gene, which provides resistance to the herbicide phosphinothricin (Thompson *et al.*, 1987). In the case of *Cattleya*, Zhang *et al.* (2010a) at Chiba University reported that the maximum transformation rate was achieved when 2 mm sectioned PLBs of *Cattleya* CM2450 (from the Orchid Sanctuary Dogashima, then Orchid Resort Dogashima, Izu, Japan, closed in 2013; grex is unknown) were first grown in NDM containing 10 g l^{-1} sucrose, 1.0 mg l^{-1} 1-naphthaleneacetic acid (NAA) and 0.1 mg l^{-1} N6-benzyladenine (BA), pH 5.4, and then inoculated with 1:10 v/v *Agrobacterium* suspension for up to 6 hours. The strain used was *Agrobacterium* EHA101 which carries the vector pIG121-Hm. This vector contains sequences for markers *nptII*, *hpt*, and *GUS*. Vector pBBRacdS was also used, which features the coding sequence for the ACC deaminase gene, whose expression decreases the ethylene level in tissues. Infected PLBs were then co-cultured for 3 days in liquid NDM supplemented with 10 mg l^{-1} of meropenem and then were transferred to solid medium supplemented with 10 mg l^{-1} hygromycin and 20 mg l^{-1} meropenem. The highest number of hygromycin-resistant PLBs per gram of inoculated PLB was obtained when they were infected for 3 hours, by infection with EHA101/pIG121. Of those, *GUS* activity was detected in 90%, and transformation was confirmed by Southern Blot. Then Zhang *et al.* (2010b) conducted a similar experiment the same year in which sectioned PLBs (2 mm as well) were co-cultivated with *Agrobacterium* strain EHA105 which carries plasmid pSMAHdN627-ORSV, and contains genes coding for *hpt* and a sequence fragment for the *Odontoglossum* ringspot virus (*ORSV*) replicase driven by the 35S promoter. They found that PLBs that were cultured in the dark, on 100 μM AS, were infected for 30 minutes and inoculated for 3 hours had the highest frequency of transformation. The highest number of hygromycin-resistant PLBs per gram of inoculated PLB was about ten. Southern blot confirmed the presence of the *ORSV* sequence, while Northern confirmed its expression.

9.4.6 *Erycina pusilla*

In comparison to popular orchids from genera like *Cymbidium*, *Dendrobium*, *Oncidium*, and *Phalaenopsis*, *Erycina pusilla* is a tiny, fast-growing species with either 10 to 12 chromosomes (Bolaños-Villegas *et al.*, 2021). As an experimental plant, *Erycina pusilla* is useful precisely because of these traits (Lin *et al.*, 2016). Until recently, reports on this orchid's transformation were scarce, but in 2015, Lee *et al.* at Cheng Kung University and Academia Sinica published an effective method for selecting and regenerating 3-month-old transfected protocorms. Strain EHA105 was used to deliver the binary vector p35S-AtMSRB7 containing the *HPT* selection marker, and the target gene *MSRB7* from *Arabidopsis* for tolerance to disease and environmental stress. Both genes were driven by the 35S promoter. The protocorms were pre-cultured for 2 days in liquid NDM supplemented with 0.1 mM AS. Then, 1 ml of *A. tumefaciens* suspension (OD600 = 0.5–0.8) was added to 30 ml of the pre-cultured protocorms. The protocorms were then immersed with the *A. tumefaciens* suspension for 4 h without shaking, and then co-cultivated (with shaking) at 24°C in the dark for 2 days until the concentration of the *A. tumefaciens* reached OD600 = 0.5. Transformed plants (*MSRB7Ox*) from the T_0 and T_1 generation were significantly tolerant to methyl viologen (MV), a bipyridine salt that induces oxidative stress (Lee *et al.*, 2015). Unlike in the wild type, detached leaves of the transgenic plants stayed green and healthy (Lee *et al.*, 2015).

9.4.7 *Vanda*

About seventy-three species make up the commercially important genus *Vanda*; including the previously accepted genera *Ascocentrum*, *Euanthe*, *Christensonia*, *Neofinetia* and *Trudelia* (Gardiner *et al.*, 2013). The genomic assembly of a Singaporean *Papilionanthe* Miss Joaquim 'Agnes' (*Papilionanthe teres* x *Papilionanthe hookeriana*) was recently finished leading to the identification of metabolic pathways involved in key floral traits (Lim *et al.*, 2022). Remarkably, less than 20 years ago, truly little was known about their genomes or their suitability for transformation (Shrestha *et al.*, 2007). Back then the development of transgenic Vandas was first attempted by Shrestha *et al.* (2007) at Chiba University on 3–4 mm PLBs (from shoot tip culture) of *Vanda* 'Tokyo Blue' by SAAT with strain EHA101/pIG121. It was determined that the best treatment was infection by 240 min and sonication for 5 min during inoculation, and successful transformation was verified by PCR and Southern Blot on seventeen plants. Then in Thailand, Pimda *et al.* (2010) transformed and regenerated plants of *V. lilacina*. They claimed to have used *Agrobacterium* strain LBA4404 with vector pCAMBIA1305.1 to deliver the coding sequence of a chitinase gene of unknown origin, for enhanced fungal tolerance. About 80% of the protocorms expressed the *GUS* marker, and transformation was confirmed by PCR only. No apparent phenotypes were reported. In Malaysia, the team of Gnasekaran *et al.* (2014) established an efficient protocol for the transfection of 12-week-old PLBs (3–4 mm) of *Vanda* Kasem's Delight Tom Boykin (*V.* Sun Tan x *V.* Thospol) with strain LBA4404 harboring plasmids pW1B1 and pW2KY for wheat pathogen resistance genes *wheatwin1* and *wheatwin 2* (*wwin1* and *wwin2*), respectively. The highest transformation rate was observed after immersion for 30 minutes in *Agrobacterium* suspension (OD600 = 0.8). Selection was made in the presence of 250 mg l^{-1} cefotaxime and 30 mg l^{-1} geneticin. PCR was done to verify the presence of the sequences for *wwin1* and *wwin2*. No phenotypes were described.

9.5 The Importance of Transformation in Resistance to Disease and Environmental Stress

Orchid breeding is important as diseases from bacteria and viruses reduce profitability. Crown rot, caused by the fungus *Phytophthora palmivora* Butl., commonly affects orchid shoot tips and can damage plants at any stage of development. Other notable fungal diseases are basal rot, attributed to *Sclerotium rolfsii* Sacc, and anthracnose, caused by *Colletotrichum gloeosporioides* (Goh, 1989), while in the case of bacteria, soft rot is caused by *Erwinia carotovora* (Ting and Ho, 1970). In subtropical regions of Asia, incidents of soft rot have been documented in *Phalaenopsis* and *Oncidium* during periods of warm and humid weather (Ting and Ho, 1971). Nonetheless, the biggest problem in orchids is viral infection, with *Odontoglossum* ringspot virus (*ORSV*) and *Cymbidium* mosaic virus (*CymMV*) being the most common causal agents (Chia *et al.* 1992; Ajjikuttira *et al.*, 2002). Unsurprisingly effective transformation strategies were developed over time in order to shield orchids against all these diseases. For instance, the *Gastrodia Antifungal Protein* (*GAFP*) and the *Neutrophils Peptide-I* (*NPI*), two genes for disease resistance, were effectively introduced into *Phalaenopsis* Tailin Angel (*P.* New Eagle x *P.* Ruey Lih Stripes) by Li *et al.* (2013) at the Nanjing Forestry University. The *NPI* gene has a wide antibiotic spectrum to bacteria, fungi, and viruses, while *GAFP* may suppress more than twenty fungi strains (Li *et al.*, 2013). The authors resorted to *A. tumefaciens* strain LBA4404 for delivery of binary vector pBin35SGAFP-NPI into PLBs regenerated from leaf segments. Following selection on medium containing 50 mg l^{-1} kanamycin and 500 mg l^{-1} cefotaxime, it was confirmed by PCR, Southern blot and RT-PCR confirmed that *GAFP* and *NPI* genes had been integrated and expressed. Finally, and *in vitro* assay with *C. gloeosporioides* suggested that these transgenic plants were disease-resistant, at least to anthracnose.

The sweet pepper ferredoxin-like protein (*pflp*) was a marker gene that more than 20 years ago was introduced to prevent infection by *Erwinia carotovora* (You *et al.*, 2003). Back then Chan *et al.* (2005) at Academia Sinica created a dual disease resistance phenotype (to viruses and bacteria) in *Phalaenopsis* PLBs by introducing sequences for: (i) the coat protein (CP) of the *pflp* and, (ii) the *Cymbidium* Mosaic Virus (CymMV). Then by allowing expression of a viral capsid protein, Chang *et al.* (2005) created stable transgenic *Dendrobium* plants resistant to CymMV. Later on, Raffeiner *et al.* (2009) induced ethylene insensitivity in flowers of transgenic *Oncidium* and *Odontoglossum* orchids by cleverly introducing the *etr1-1* allele from *Arabidopsis thaliana*. Finally, a gene encoding a lipid transfer protein was introduced into *Phalaenopsis amabilis* by Qin *et al.* (2011) to increase resistance to cold.

9.6 Conclusions

Advances in transformation technologies have led to the commercial availability of transgenic crops. Biotechnological methods could help develop effective transformation processes for orchids, providing them with unique traits. Crucial factors for success include bacterial strains, reporter genes, selectable markers, PLB selection periods, and culture conditions. The main target tissue for transformation with *A. tumefaciens* are still the PLBs. However, studies on the selection processes done after infection vary a lot. Because extended culture in non-selection media may lead to chimeric sectors or escape plants, most reports describe slow selection methods lasting from one week to several months. Fortunately, several genome initiatives focused on orchids are currently in progress, aiming to identify and validate the function of many genes. These studies will without a doubt bring about highly effective and replicable transformation processes, thus opening an exciting new era in orchid breeding for the benefit of growers and consumers across the world.

References

Ajjikuttira, P.A, Lim-Ho, C.L., Woon, M.H., Ryu, K.H., Chang, C.A., Loh, C.S. and Wong, S.M. (2002) Genetic variability in the coat protein genes of two orchid viruses: *Cymbidium* mosaic virus and *Odontoglossum* ringspot virus. *Archives of Virology* 147, 1943–1954. DOI: 10.1007/s00705-002-0868-5

Anzai, H. and Tanaka, M. (2001) Transgenic *Phalaenopsis* (a moth orchid). In: Bajaj, Y.P.S (ed.) *Transgenic Crops III. Biotechnology in Agriculture and Forestry, vol 48.* Springer, Berlin, Germany, pp. 249–264. https://doi.org/10.1007/978-3-662-10603-7_18 (accessed 10 July 2025).

Anzai, H., IsHII, Y., Shichinohe, M., Katsumata, K., Nojiri, C., Morikawa, H. and Tanaka, M. (1996) Transformation of *Phalaenopsis* by particle bombardment. *Plant Tissue Culture Letters* 13, 265–272. DOI: 10.5511/plantbiotechnology1984.13.265

Arditti, J. (1992) *Fundamentals of Orchid Biology*. John Wiley and Sons, New York, New York.

Arditti, J. and Ernst, R. (1993) *Micropropagation of Orchids*. John Wiley, New York, New York.

Azadi, P., Otang, N.V., Chin, D.P., Nakamura, I., Fujisawa, M., Harada, H., Misawa, N. and Mii, M. (2010) Metabolic engineering of *Lilium x formolongi* using multiple genes of the carotenoid biosynthesis pathway. *Plant Biotechnology Reports* 4, 269–280. DOI: 10.1007/s11816-010-0147-y

Belarmino, M.M. and Mii, M. (2000) *Agrobacterium*-mediated genetic transformation of a *Phalaenopsis* orchid. *Plant Cell Reports* 19, 435–442. DOI: 10.1007/s002990050752

Bhattacharyya, P., Kumaria, S., Diengdoh, R. and Tandon, P. (2014) Genetic stability and phytochemical analysis of the *in vitro* regenerated plants of *Dendrobium nobile* Lindl., an endangered medicinal orchid. *Meta Gene* 2, 489–504. DOI: 10.1016/j.mgene.2014.06.003

Bidney, D., Scelonge, C., Martich, J., Burrus, M., Sims, L. and Huffman, G. (1992) Microprojectile bombardment of plant tissues increases transformation frequency by *Agrobacterium tumefaciens*. *Plant Molecular Biology* 18, 301–313.

Bolanos-Villegas, P. and Chen F.C. (2022) Advances and perspectives for polyploidy breeding in orchids. *Plants* 11, 1421. DOI: 10.3390/plants11111421

Bolaños-Villegas, P., Chang, C., and Chen, F. C. (2021) The tiny twig epiphyte *Erycina pusilla*, a model for orchid genome and breeding research. In: Chen, F.C. and Chin, S.W. (eds.) *The Orchid Genome, Compendium of Plant Genomes*. Springer Nature, Cham, Switzerland, pp. 41–47. DOI: 10.1007/978-3-030-66826-6_3

Bolton, E.E., Chen, J., Kim, S. *et al.* (2011). PubChem3D: a new resource for scientists. *Journal of Cheminformatics* 3, 32. DOI: 10.1186/1758-2946-3-32

Brasileiro A.C.M., Aragao F.J.L., Rossi S., Dusi D.M.A., Gomes Barros L.M. and Rech, E.L. (1996) Susceptibility of common and tepary beans to *Agrobacterium* spp. strains and improvement of *Agrobacterium*-mediated transformation using microprojectile bombardment. *Journal of the American Society of Horticultural Science* 121, 810–815.

Cardoso J.C., Zanello C.A. and Chen, J.T. (2020) An overview of orchid protocorm-like bodies: mass propagation, biotechnology, molecular aspects, and breeding. *International Journal of Molecular Sciences*, 21, 985. DOI: 10.3390/ijms21030985

Chai, D. and Yu, H. (2007) Recent advances in transgenic orchid production. *Orchid Science and Biotechnology* 1, 34–39.

Chai, M.L., Xu, C.J., Senthil, K.K., Kim, J.Y. and Kim, D.H. (2002) Stable transformation of protocorm-like bodies in *Phalaenopsis* orchid mediated by *Agrobacterium tumefaciens*. *Scientia Horticulturae* 96, 213–224. DOI: 10.1016/S0304-4238(02)00084-5

Chan, Y.L, Chen, W.H. and Chan, M.T. (2003) *Phalaenopsis* gene transformation (I) - optimization of transient gene expression (original in Chinese). *Journal of the Chinese Society of Horticultural Science* (中國園藝) 49, 33–44. DOI: 10.6964/JCSHS.200303.0033

Chan, Y.L., Lin, K.H., Sanjaya, Liao, L.J., Chen, W.H. and Chan, M.T. (2005) Gene stacking in *Phalaenopsis* orchid enhances dual tolerance to pathogen attack. *Transgenic Research* 14, 279–288. DOI: 10.1007/s11248-005-0106-5

Chang, C., Chen, Y.C., Hsu, Y.H., Wu, J.T., Hu, C.C., Chang, W.C. and Lin, N.S. (2005) Transgenic resistance to *Cymbidium* mosaic virus in *Dendrobium* expressing the viral capsid protein gene. *Transgenic Research* 14, 41–46. DOI: 10.1007/s11248-004-2373-y

Cheng, Y.H., Yang, J.S. and Yeh, S.D. (1996) Efficient transformation of papaya by coat protein gene of papaya ringspot virus mediated by *Agrobacterium* following liquid-phase wounding of embryogenic tissues with carborundum. *Plant Cell Reports* 16, 127–132. DOI: 10.1007/BF01890852

Chia T.F., Chan Y.S. and Chua, N.H. (1992) Detection and localization of viruses in orchids by tissue-print hybridization. *Plant Pathology* 41, 355–361.

Chia T.F., Chan Y.S. and Chua N.H. (1994) The firefly luciferase gene as a non-invasive reporter for *Dendrobium* transformation. *The Plant Journal* 6, 441–446.

Chia, T.F., Lim, A.Y.H., Luan, Y. and Ng, I. (2001). Transgenic *Dendrobium* (orchid). In: Bajaj, Y.P.S. (ed.) *Transgenic crops III. Biotechnology in agriculture and forestry, vol 48*. Springer, Berlin, Germany, pp. 95–106. DOI: 10.1007/978-3-662-10603-7_8

Chin, D.P., Mishiba, K.I. and Mii, M. (2007) *Agrobacterium*-mediated transformation of protocorm-like bodies in *Cymbidium*. *Plant Cell Reports* 26, 735–743. DOI: 10.1007/s00299-006-0284-5

Chinsamy, M., Finnie, J.F. and Van Staden, J. (2014) Anti-inflammatory, antioxidant, anti-cholinesterase activity and mutagenicity of South African medicinal orchids. *South African Journal of Botany* 91, 88–98.

Chugh, S., Guha, S. and Rao, I.U. (2009) Micropropagation of orchids: a review on the potential of different explants. *Scientia Horticulturae* 122, 507–520. DOI: 10.1016/j.scienta.2009.07.016

De Cleene, M. and De Ley, J. (1976) The host range of crown gall. *The Botanical Review* 42, 389–466. DOI: 10.1007/BF02860827

Ding, L., Wang, Y. and Yu, H. (2013) Overexpression of *DOSOC1*, an ortholog of Arabidopsis *SOC1*, promotes flowering in the orchid *Dendrobium* Chao Parya Smile. *Plant Cell Physiology* 54, 595–608. DOI: 10.1093/pcp/pct026

Dressler, R.L. (2005) How many orchid species? *Selbyana* 26, 155–158. https://www.jstor.org/stable/41760186 (accessed 10 July 2025).

Fang, Y.D., Akula, C. and Altpeter, F. (2002) *Agrobacterium*-mediated barley (*Hordeum vulgare* L.) transformation using green fluorescent protein as a visual marker and sequence analysis of the T-DNA∝ barley genomic DNA junctions. *Journal of Plant Physiology* 159, 1131–1138. DOI: 10.1078/0176-1617-00707

Fu, C.H., Chen, Y.W., Hsiao, Y.Y., Pan, Z.J., Liu, Z.J., Huang, Y.M., Tsai, W.C. and Chen, H.H. (2011) OrchidBase: a collection of sequences of the transcriptome derived from orchids. *Plant and Cell Physiology* 52, 238–243. DOI: 10.1093/pcp/pcq201

Gardiner, L.M., Kocyan, A., Motes, M., Roberts, D.L. and Emerson, B.C. (2013) Molecular phylogenetics of *Vanda* and related genera (Orchidaceae). *Botanical Journal of the Linnean Society* 173, 549–572. DOI: 10.1111/boj.12102

Gnasekaran, P., James Antony, J.J., Uddain, J. and Subramaniam, S. (2014) *Agrobacterium*-mediated transformation of the recalcitrant *Vanda* Kasem's Delight orchid with higher efficiency. *The Scientific World Journal* 2014, 583934. DOI: 10.1155/2014/583934

Goh, C.J. (1989) Orchids, monopodials. In: Ammirato, P.V., Evans, D.R., Sharp, W.R. and Bajaj, Y.P.S. (eds) *Handbook of Plant Cell Culture: Ornamental Species Vol. 5*. Macmillan and McGraw-Hill, New York, New York, pp. 598–653.

Gordon-Kamm, W.J., Spencer, T.M., Mangano, M.L., Adams, T.R., Daines, R.J. *et al.* (1990) Transformation of maize cells and regeneration of fertile transgenic plants. *The Plant Cell* 2, 603–618. DOI: 10.1105/tpc.2.7.603

Govaerts, R., Nic Lughadha, E., Black, N., Turner, R. and Paton, A. (2021) The world checklist of vascular plants, a continuously updated resource for exploring global plant diversity. *Scientific Data* 8, 215. DOI: 10.1038/s41597-021-00997-6

Grayburn, W.S. and Vick, B.A. (1995) Transformation of sunflower (*Helianthus annuus* L.) following wounding with glass beads. *Plant Cell Reports* 14, 285–289. DOI: 10.1007/BF00232029

Gurel, S., Gurel, E., Kaur, R., Wong, J., Meng, L., Tan, H.Q. and Lemaux, P.G. (2009) Efficient, reproducible *Agrobacterium*-mediated transformation of sorghum using heat treatment of immature embryos. *Plant Cell Reports* 28, 429–444. DOI: 10.1007/s00299-008-0655-1

Gutiérrez, R.M.P. (2010). Orchids: a review of uses in traditional medicine, its phytochemistry and pharmacology. *Journal of Medicinal Plants Research* 4, 592–638. http://www.academicjournals.org/jmpr/PDF/pdf2010/18Apr/Gutierrez.pdf (accessed 10 July 2025).

Hiei, Y., Ishida, Y. and Komari, T. (2014) Progress of cereal transformation technology mediated by *Agrobacterium tumefaciens*. *Frontiers in Plant Science* 5, 628. DOI: 10.3389/fpls.2014.00628

Hisano, H., Meints, B., Moscou, M.J., Cistue, L., Echávarri, B., Sato, K. and Hayes, P.M. (2017) Selection of transformation-efficient barley genotypes based on TFA (transformation amenability) haplotype and higher resolution mapping of the *TFA* loci. *Plant Cell Reports* 36, 611–620. DOI: 10.1007/s00299-017-2107-2

Hoshi, Y., Kondo, M., Mori, S., Adachi, Y., Nakano, M. and Kobayashi, H. (2004). Production of transgenic lily plants by *Agrobacterium*-mediated transformation. *Plant Cell Reports* 22, 359–364. DOI: 10.1007/s00299-003-0700-z

Hsiao, Y.Y., Pan, Z.J., Hsu, C.C., Yang, Y.P., Hsu, Y.C. *et al.* (2011) Research on orchid biology and biotechnology. *Plant and Cell Physiology* 52, 1467–1486. DOI: 10.1093/pcp/pcr100 (accessed 10 July 2025).

Hsieh Y.H., Hsu T.Y. and Huang P.L. (1995) Studies on genetic transformation of *Phalaenopsis* by microprojectile bombardment (original in Chinese). *Journal of the Chinese Society of Horticultural Science* (中國園藝) 41, 174–185.

Hsing, H.X., Lin, Y.J., Tong, C.G., Li, M.J., Chen, Y.J. and Ko, S.S. (2016) Efficient and heritable transformation of *Phalaenopsis* orchids. *Botanical Studies* 57, 1–12. DOI: 10.1186/s40529-016-0146-6

Hu, W., Li, W., Xie, S., Fagundez, S., McAvoy, R., Deng, Z. and Li, Y. (2016) *Kn1* gene overexpression drastically improves genetic transformation efficiencies of citrus cultivars. *Plant Cell, Tissue and Organ Culture* 125, 81–91. DOI: 10.1007/s11240-015-0931-z

Huang, J.Z., Lin, C.P., Cheng, T.C., Chang, B.C.H., Cheng, S.Y. *et al.* (2015) A *de novo* floral transcriptome reveals clues into *Phalaenopsis* orchid flower development. *PLOS ONE* 10, p.e0123474. DOI: 10.1371/journal.pone.0123474 (accessed 10 July 2025).

Huang, Y.W., Tsai, Y.J., Cheng, T.C., Chen, J.J. and Chen, F.C. (2014) Physical wounding and ethylene stimulated embryogenic stem cell proliferation and plantlet regeneration in protocorm-like bodies of *Phalaenopsis* orchids. *Genetics and Molecular Research* 13, 9543–9557. DOI: 10.4238/2014.November.12.3.

Huang, J.Z., Lin, C.P., Cheng, T.C., Huang, Y.W., Tsai, Y.J. *et al.* (2016) The genome and transcriptome of *Phalaenopsis* yield insights into floral organ development and flowering regulation. *PeerJ* 4, p.e2017. DOI: 10.7717/peerj.2017

Ishida, Y., Hiei, Y. and Komari, T. (2007) *Agrobacterium*-mediated transformation of maize. *Nature Protocols* 2, 1614–1621. DOI: 10.1038/nprot.2007.241

Knapp, J.E., Kausch, A.P. and Chandlee, J.M. (2000) Transformation of three genera of orchid using the bar gene as a selectable marker. *Plant Cell Reports* 19, 893–898. DOI: 10.1007/s002990000202

Knittel, N., Gruber, V., Hahne, G. and Lénée, P. (1994) Transformation of sunflower (*Helianthus annuus* L.): a reliable protocol. *Plant Cell Reports* 14, 81–86. DOI: 10.1007/BF00233766

Kui, L., Chen, H., Zhang, W., He, S., Xiong, Z., Zhang, Y., Yan, L., Zhong, C., He, F., Chen, J. and Zeng, P. (2017) Building a genetic manipulation tool box for orchid biology: identification of constitutive promoters and application of CRISPR/Cas9 in the orchid, *Dendrobium officinale*. *Frontiers in Plant Science* 7, 2036. DOI: 10.3389/fpls.2016.02036

Laws, N. (1995) Cut orchids in the world market. *FloraCulture International* 5, 12–15.

Lee, S.H., Li, C.W., Liau, C.H., Chang, P.Y., Liao, L.J., Lin, C.S. and Chan, M.T. (2015) Establishment of an *Agrobacterium*-mediated genetic transformation procedure for the experimental model orchid *Erycina pusilla*. *Plant Cell, Tissue and Organ Culture* 120, 211–220. DOI: 10.1007/s11240-014-0596-z

Lee, J.Y. (2023) The principles and applications of high-throughput sequencing technologies. *Development and Reproduction* 27, 9. DOI: 10.12717/DR.2023.27.1.9

Leitch, I.J., Kahandawala, I., Suda, J., Hanson, L., Ingrouille, M.J., Chase, M.W. and Fay, M.F. (2009) Genome size diversity in orchids: consequences and evolution. *Annals of Botany* 104, 469–481. DOI: 10.1093/aob/mcp003

Li, S.H., Kuoh, C.S., Chen, Y.H., Chen, H.H. and Chen, W.H. (2005) Osmotic sucrose enhancement of single-cell embryogenesis and transformation efficiency in *Oncidium*. *Plant Cell, Tissue and Organ Culture* 81, 183–192. DOI: 10.1007/s11240-004-4955-z

Li, J., Kuang, P., Liu, R.D., Wang, D., Wang, Z.N. and Huang, M.R. (2013) Transfer of the *GAFP* and *NPI*, two disease-resistant genes, into a *Phalaenopsis* by *Agrobacterium tumefaciens*. *Pakistan Journal of Botany*, 45, 1761–1766.

Liao, Y.J., Tsai, Y.C., Sun, Y.W., Lin, R.S. and Wu, F.S. (2011) *In vitro* shoot induction and plant regeneration from flower buds in *Paphiopedilum* orchids. *In Vitro Cellular and Developmental Biology-Plant* 47, 702–709. DOI: 10.1007/s11627-011-9370–9377

Liau, C.H., Lu, J.C., Prasad, V., Hsiao, H.H., You, S.J. *et al.* (2003a) The sweet pepper ferredoxin-like protein (*pflp*) conferred resistance against soft rot disease in *Oncidium* orchid. *Transgenic Research* 12, 29–336. DOI: 10.1023/A:1023343620729

Liau, C.H., You, S.J., Prasad, V., Hsiao, H.H., Lu, J.C., Yang, N.S. and Chan, M.T. (2003b) *Agrobacterium tumefaciens*-mediated transformation of an *Oncidium* orchid. *Plant Cell Reports* 21, 993–998. DOI: 10.1007/s00299-003-0614-9

Lim, A.H., Low, Z.J., Shingate, P.N., Hong, J.H., Chong, S.C. *et al.* (2022) Genome assembly and chemogenomic profiling of National Flower of Singapore *Papilionanthe* Miss Joaquim 'Agnes' reveals metabolic pathways regulating floral traits. *Communications Biology* 5, 967. DOI: 10.1038/s42003-022-03940-6

Lin, C.S., Hsu, C.T., Liao, D.C., Chang, W.J., Chou, M.L. *et al.* (2016) Transcriptome-wide analysis of the MADS-box gene family in the orchid *Erycina pusilla*. *Plant Biotechnology Journal* 14, 284–298. DOI: 10.1111/pbi.12383

Lipsky, A., Cohen, A., Gaba, V., Kamo, K., Gera, A. and Watad, A. (2000) Transformation of *Lilium longiflorum* plants for cucumber mosaic virus resistance by particle bombardment. *Acta Horticulturae* 568, 30. DOI:10.17660/ActaHortic.2002.568.30

Liu, J.X., Chiou, C.Y., Shen, C.H., Chen, P.J., Liu, Y.C. *et al.* (2014) RNA interference-based gene silencing of phytoene synthase impairs growth, carotenoids, and plastid phenotype in *Oncidium* hybrid orchid. *SpringerPlus* 3, 1–13. DOI: 10.1186/2193-1801-3-478

Men, S., Ming, X., Liu, R., Wei, C. and Li, Y. (2003a) *Agrobacterium*-mediated genetic transformation of a *Dendrobium* orchid. *Plant Cell, Tissue and Organ Culture* 75, 63–71. DOI: 10.1023/A:1024627917470

Men, S., Ming, X., Wang, Y., Liu, R., Wei, C. and Li, Y. (2003b) Genetic transformation of two species of orchid by biolistic bombardment. *Plant Cell, Tissue and Organ Culture* 21, 592–598. DOI: 10.1023/A:1024627917470

Mii, M. and Chin, D.P. (2010) Genetic transformation of orchids. *Acta Horticulturae* 878, 461–466. DOI: 10.17660/ActaHortic.2010.878.59

Mishiba, K., Chin, D.P. and Mii, M. (2005) *Agrobacterium*-mediated transformation of *Phalaenopsis* by targeting protocorms at an early stage after germination. *Plant Cell Reports* 24, 297–303. DOI: 10.1007/s00299-005-0938-8

Montoro, P., Rattana, W., Pujade-Renaud, V., Michaux-Ferrière, N., Monkolsook, Y., Kanthapura, R. and Adunsadthapong, S. (2003) Production *of Hevea brasiliensis* embryogenic callus lines by *Agrobacterium tumefaciens*: roles of calcium. *Plant Cell Reports* 21, 1065–1102. DOI: 10.1007/s00299-003-0632-7

Nan, G.L (1998) Transgenic *Dendrobium* orchid through *Agrobacterium*-mediated transformation. *Malay Orchid Review* 32, 93–96.

Nan, G.L. and Kuehnle, A.R. (1995) Factors affecting gene delivery by particle bombardment of *Dendrobium* orchids. *In Vitro Cellular and Developmental Biology-Plant* 31, 131–136. DOI: 10.1007/BF02632007

Nan, G.L., Tang, C.S., Kuehnle, A.R. and Kado, C.I. (1997) *Dendrobium* orchids contain an inducer of *Agrobacterium* virulence genes. *Physiological and Molecular Plant Pathology* 51, 391–399. DOI: 10.1006/pmpp.1997.0128

O'Neill, S.D. (1997) Pollination regulation of flower development. *Annual Review of Plant Biology* 48, 547–574. DOI: 10.1146/annurev.arplant.48.1.547

Olhoft, P., Lin, K., Galbraith, J., Nielsen, N., and Somers, D. (2001) The role of thiol compounds in increasing *Agrobacterium* mediated transformation of soybean cotyledonary-node cells. *Plant Cell Reports* 20, 731–737. DOI: 10.1007/s002990100388

Patel, M., Dewey, R.E. and Qu, R. (2013) Enhancing *Agrobacterium tumefaciens*-mediated transformation efficiency of perennial ryegrass and rice using heat and high maltose treatments during bacterial infection. *Plant Cell, Tissue and Organ Culture* 114, 19–29. DOI: 10.1007/s11240-013-0301-7

Phlaetita, W., Chin, D.P., Otang, N.V., Nakamura, I. and Mii, M. (2015) High efficiency Agrobacterium-mediated transformation of *Dendrobium* orchid using protocorms as a target material. *Plant Biotechnology* 32, 323–327. DOI: 10.5511/plantbiotechnology.15.0804a

Pimda, W. and Bunnag, S. (2010) Genetic transformation of *Vanda lilacina* Teijsm. & Binnend. with a chitinase gene. *Advances in Agriculture and Botanic* 2, 71–78.

Potrykus, I. (1991) Gene transfer to plants: assessment of published approaches and results. *Annual Review of Plant Physiology and Plant Molecular Biology* 42, 205–225. DOI: 10.1146/annurev.pp.42.060191.001225

Qin, X., Liu, Y., Mao, S., Li, T., Wu, H., Chu, C. and Wang, Y. (2011) Genetic transformation of lipid transfer protein encoding gene in *Phalaenopsis amabilis* to enhance cold resistance. *Euphytica* 177, 33–43. DOI: 10.1007/s10681-010-0246-4

Quan, Z., Yongping, Z., Guangdong, W., Weiming, G. and Zhiguo, Z. (2011) Sonication assisted *Agrobacterium*-mediated transformation of chalcone synthase (CHS) gene to spring *Dendrobium* cultivar 'Sanya'. *African Journal of Biotechnology* 10, 11832–11838. DOI: 10.5897/AJB11.224

Raffeiner, B., Serek, M. and Winkelmann, T. (2009) *Agrobacterium tumefaciens*-mediated transformation of *Oncidium* and *Odontoglossum* orchid species with the ethylene receptor mutant gene *etr1-1*. *Plant Cell, Tissue and Organ Culture* 98, 125–134. DOI: 10.1007/s11240-009-9545-7

Roberts, R.L., Metz, M., Monks, D.E., Mullaney, M.L., Hall, T. and Nester, E.W. (2003) Purine synthesis and increased *Agrobacterium tumefaciens* transformation of yeast and plants. *Proceedings of the National Academy of Sciences of the United States of America* 100, 6634–6639. DOI: 10.1073/pnas.113202210

Saika, H., and Toki, S. (2010) Mature seed-derived callus of the model *indica* rice variety Kasalath is highly competent in *Agrobacterium*-mediated transformation. *Plant Cell Reports* 29, 1351–1364. DOI: 10.1007/s00299-010-0921-x

Sanjaya, and Chan, M.T. (2007) Genetic transformation as a tool for improvement of orchids. In: Chen, W.H. and Chen, H.H. (eds) *Orchid Biotechnology*. World Scientific Publishing, Singapore, pp. 225–253.

Sawettalake, N., Bunnag, S., Wang, Y., Shen, L. and Yu, H. (2017) *DOAP1* promotes flowering in the orchid *Dendrobium* Chao Praya Smile. *Frontiers in Plant Science* 8, 400. DOI: 10.3389/fpls.2017.00400

Semiarti, E., Indrianto, A., Purwantoro, A., Isminingsih, S., Suseno, N., Ishikawa, T., Yoshioka, Y., Machida, Y. and Machida, C. (2007) *Agrobacterium*-mediated transformation of the wild orchid species *Phalaenopsis amabilis*. *Plant Biotechnology* 24, 265–272. DOI: 10.5511/plantbiotechnology.24.265

Semiarti, E., Purwantoro, A., Mercuriani, I.S., Anggriasari, A.M., Jang, S., Suhandono, S., Machida, Y. and Machida, C. (2014) *In planta* transformation method for T-DNA transfer in orchids. *Proceedings of the 4th International Conference on Mathematics and Natural Sciences* 1589, 303–307. http://dx.doi.org/10.1063/1.4868806

Serna-Sánchez, M.A., Pérez-Escobar, O.A., Bogarín, D., Torres-Jiménez, M.F., Álvarez-Yela, A.C., Arcila-Galvis, J.E., Hall, C.F., De Barros, F., Pinheiro, F., Dodsworth, S. and Chase, M.W. (2021) Plastid phylogenomics resolves ambiguous relationships within the orchid family and provides a solid timeframe for biogeography and macroevolution. *Scientific Reports* 11, 6858. DOI: 10.1038/s41598-021-83664-5

Shou, H., Frame, B.R., Whitham, S.A. and Wang, K. (2004) Assessment of transgenic maize events produced by particle bombardment or *Agrobacterium*-mediated transformation. *Molecular Breeding* 13, 201–208. DOI: 10.1023/B:MOLB.0000018767.64586.53

Shrestha, B.R., Chin, D.P., Tokuhara, K. and Mii, M. (2007) Efficient production of transgenic plants of *Vanda* through sonication-assisted *Agrobacterium*-mediated transformation of protocorm-like bodies. *Plant Biotechnology* 24, 429–434. DOI: 10.5511/plantbiotechnology.24.429

Someya, T., Nonaka, S., Nakamura, K. and Ezura, H. (2013) Increased 1-aminocyclopropane-1-carboxylate deaminase activity enhances *Agrobacterium tumefaciens*-mediated gene delivery into plant cells. *MicrobiologyOpen* 2, 873–880. DOI: 10.1002/mbo3.123

Stachel, S.E., Messens, E., Van Montagu, M. and Zambryski, P. (1985). Identification of the signal molecules produced by wounded plant cells that activate T-DNA transfer in *Agrobacterium tumefaciens*. *Nature* 318, 624–629. DOI: 10.1038/318624a0

Su, C.L., Chao, Y.T., Yen, S.H., Chen, C.Y., Chen, W.C., Chang, Y.C.A. and Shih, M.C. (2013) Orchidstra: an integrated orchid functional genomics database. *Plant and Cell Physiology* 54, e11. DOI: 10.1093/pcp/pct004

Subramaniam, S., and Xavier, R. (2010) Emerging factors that influence efficiency of T-DNA gene transfer into *Phalaenopsis violacea* orchid via *Agrobacterium tumefaciens*-mediated transformation system. *International Journal of Biology* 2, 64–73. http://www.ccsenet.org/journal/index.php/ijb/article/view/4415/5204

Subramaniam, S., Samian, R., Midrarullah and Rathinam, X. (2009) Preliminary factors influencing transient expression of *Gusa* in *Dendrobium* Savin White protocorm-like bodies (PLBs) using *Agrobacterium*-mediated transformation system. *World Applied Sciences Journal* 7, 1295–1307.

Subramaniam, S., and Rahman, Z. A. (2010) Early GFP gene assessments influencing *Agrobacterium tumefaciens*-mediated transformation system in *Phalaenopsis violacea* orchid. *Emirates Journal of Food and Agriculture* 22, 103. https://tinyurl.com/mrxat4b3 (accessed 10 July 2025).

Sundar, I.K. and Sakthivel, N. (2008) Advances in selectable marker genes for plant transformation. *Journal of Plant Physiology* 165, 1698–1716. DOI: 10.1016/j.jplph.2008.08.002.

Teixeira da Silva, J.A (2013) Orchids: advances in tissue culture, genetics, phytochemistry and transgenic biotechnology. *Floriculture and Ornamental Biotechnology* 7, 1–52.

Teixeira da Silva, J.A, Chin, D.P., Van, P.T. and Mii, M. (2011) Transgenic orchids. *Scientia Horticulturae* 130, 673–680. DOI: 10.1016/j.scienta.2011.08.025

Thiruvengadam, M., Hsu, W.H., and Yang, C.H. (2011) Phosphomannose-isomerase as a selectable marker to recover transgenic orchid plants (*Oncidium* Gower Ramsey). *Plant Cell, Tissue and Organ Culture* 104, 239–246. DOI: 10.1007/s11240-010-9827-0

Thiruvengadam, M., Chung, I.M., and Yang, C.H. (2012) Overexpression of *Oncidium* MADS box (*OMADS1*) gene promotes early flowering in transgenic orchid (*Oncidium* Gower Ramsey). *Acta Physiologiae Plantarum* 34, 1295–1302. DOI: 10.1007/s11738-012-0926-x

Thompson, C.J., Movva, N.R., Tizard, R., Crameri, R., Davies, J. E., Lauwereys, M. and Botterman, J. (1987) Characterization of the herbicide-resistance gene *bar* from *Streptomyces hygroscopicus*. *The EMBO Journal* 6, 2519–2523. DOI: 10.1002/j.1460-2075.1987.tb02538.x

Ting, W.P. and Ho, T.H. (1971) *Disease and pests of orchids*. Ministry of Agriculture and Lands, Kuala Lumpur.

Tiwari, P., Sharma, A., Bose, S.K. and Park, K.I. (2024) Advances in orchid biology: biotechnological achievements, translational success, and commercial outcomes. *Horticulturae* 10, 152. DOI: 10.3390/horticulturae10020152

Tokuhara, K. and Mii, M. (1993) Micropropagation of *Phalaenopsis* and *Doritaenopsis* by shoot tips of flower stalk buds. *Plant Cell Reports* 13, 7–11. DOI: 10.1007/BF00232306

Tokuhara, K. and Mii, M. (2001) Induction of embryogenic callus and cell suspension culture from shoot tips excised from flower stalk buds of *Phalaenopsis* (Orchidaceae). *In Vitro Cellular and Developmental Biology-Plant* 37, 457–461. DOI: 10.1007/s11627-001-0080-4

Tsay, H.S., Ho, H.M., Gupta, S.K., Wang, C.S., Chen, P.T. and Chen, E.C.F. (2012) Development of pollen mediated activation tagging system for *Phalaenopsis* and *Doritaenopsis*. *Electronic Journal of Biotechnology* 15, 9. DOI: 10.2225/vol15-issue4-fulltext-1

Uddain, J., Zakaria, L., Lynn, C.B. and Subramaniam, S. (2015) Preliminary assessment on *Agrobacterium*-mediated transformation of *Dendrobium* Broga Giant orchid's PLBs. *Emirates Journal of Food and Agriculture* 27, 669. DOI: 10.9755/ejfa.2015.05.211

Vitt, P., Taylor, A., Rakosy, D., Kreft, H., Meyer, A., Weigelt, P. and Knight, T.M. (2023) Global conservation prioritization for the Orchidaceae. *Scientific Reports* 13, 6718. DOI: 10.1038/s41598-023-30177-y

Wang, W., Ling, Y.X., Yang, J., Cheng, Z., Zhu, Y.G., and Li, S. (2011) Study on *Agrobacterium tumefaciens*-mediated transformation of *Cymbidium hybridum* (original text in Chinese). *Acta Botanica Boreali-Occidentalia Sinica* (西北植物学报) 31, 27–32. https://www.cabidigitallibrary.org/doi/full/10.5555/20113069162 (accessed 10 July 2025).

Wang, S.L., Viswanath, K.K., Tong, C.G., An, H.R., Jang, S. and Chen, F.C. (2019) Floral induction and flower development of orchids. *Frontiers in Plant Science* 10, 1258. DOI: 10.3389/fpls.2019.01258

Wu, E., Lenderts, B., Glassman, K., Berezowska-Kaniewska, M., Christensen, H. *et al.* (2014) Optimized *Agrobacterium*-mediated sorghum transformation protocol and molecular data of transgenic sorghum plants. *In Vitro Cellular and Developmental Biology-Plant* 50, 9–18. DOI: 10.1007/s11627-013-9583-z

Xie, L., Wang, F., Zeng, R., Guo, H., Zhou, Y. and Zhang, Z. (2015). *Agrobacterium*-mediated transformation of *Cymbidium sinensis*. *Chinese Journal of Biotechnology* (生物工程学报) 31, 542–551. DOI: 10.13345/j.cjb.140358

Yang, J., Lee, H.J., Shin, D.H., Oh, S.K., Seon, J.H., Paek, K.Y. and Han, K.H. (1999) Genetic transformation of *Cymbidium* orchid by particle bombardment. *Plant Cell Reports* 18, 978–984. DOI: 10.1007/s002990050694

Yan, L., Wang, X., Liu, H., Tian, Y., Lian, J., Yang, R., Hao, S., Wang, X., Yang, S., Li, Q. and Qi, S. (2015) The genome of *Dendrobium officinale* illuminates the biology of the important traditional Chinese orchid herb. *Molecular Plant* 8, 922–934. DOI: 10.1016/j.molp.2014.12.011

Yeung, E.C. (2022) The orchid embryo - "an embryonic protocorm". *Botany* 100, 691–706. DOI: 10.1139/cjb-2022-0017

You, S.J., Liau, C.H., Huang, H.E., Feng, T.Y., Prasad, V. *et al.* (2003) Sweet pepper ferredoxin-like protein (*pflp*) gene as a novel selection marker for orchid transformation. *Planta* 217, 60–65. DOI: 10.1007/s00425-002-0970-7

Yu, Z., Chen, M., Nie, L., Lu, H., Ming, X., Zheng, H., Qu, L.J. and Chen, Z. (1999) Recovery of transgenic orchid plants with hygromycin selection by particle bombardment to protocorms. *Plant Cell, Tissue and Organ Culture* 58, 87–92. DOI: 10.1023/A:1006334605947

Yu, H., Yang, S.H. and Goh, C.J. (2001) *Agrobacterium*-mediated transformation of a *Dendrobium* orchid with the class 1 *knox* gene *DOH1*. *Plant Cell Reports* 20, 301–305. DOI: 10.1007/s002990100334

Zambre, M., Terryn, N., De Clercq, J., De Buck, S., Dillen, W., Van Montagu, M., Van Der Straeten, D. and Angenon, G. (2003) Light strongly promotes gene transfer from *Agrobacterium tumefaciens* to plant cells. *Planta* 216, 580–586. DOI: 10.1007/s00425-002-0914-2

Zhang, X.S. and O'Neill, S.D. (1993) Ovary and gametophyte development are coordinately regulated by auxin and ethylene following pollination. *The Plant Cell*, 5, 403–418. DOI: 10.1105/tpc.5.4.403

Zhang, L., Chin, D.P. and Mii, M. (2010a) *Agrobacterium*-mediated transformation of protocorm-like bodies in *Cattleya*. *Plant Cell, Tissue and Organ Culture* 103, 41–47. DOI: 10.1007/s11240-010-9751-3

Zhang, L., Chin, D. P., Fukami, M., Ichikawa, H., Nakamura, I., and Mii, M. (2010b) *Agrobacterium*-mediated genetic transformation of *Cattleya* with an *Odontoglossum* ringspot virus replicase gene sequence. *Plant Biotechnology* 27, 421–426. DOI: 10.5511/plantbiotechnology.10.0720a

Zion Market Research (2024) Orchid market size, share, trends, analysis. Available at: www.zionmarketresearch.com/report/orchid-market (accessed 18 July 2025).

10 Biotechnology and Breeding of Orchids in Brazil

Jean Carlos Cardoso*

Laboratory of Plant Physiology and Tissue Culture, Department of Biotechnology, Plant and Animal Production, Center for Agricultural Sciences, Federal University of Sao Carlos, Araras, São Paulo, Brazil

Abstract

The orchid family contains many species that are traded or may have the potential to be traded as ornamentals. The genera *Phalaenopsis*, *Dendrobium*, *Cymbidium*, *Oncidium*, and *Cattleya* constitute most of those plants sold worldwide. Most of the orchids used for this purpose are interspecific and intergeneric hybrids of the aforementioned genera. The breeding of these groups predominantly occurs through controlled manual hybridization, employing biotechnology for *in vitro* cultivation, germination, and initial seedling development using asymbiotic germination in a culture medium under controlled conditions. The resulting progenies are acclimatized and grown in a greenhouse to select superior phenotypes that interest national and international floricultural businesses. Moreover, *in vitro* plant culture rapidly multiplies clonal plantlets and ensures high phytosanitary quality, facilitating the exchange, shipment, and sale of new flower hybrids in the world market. Thus, biotechnology will continue to play a key role in improving orchid genetics in Brazil and ensuring the supply of high-quality micropropagated plantlets to the flower market.

Keywords: Brazilian orchid market, conventional hybridization, asymbiotic germination, clonal plantlet production, orchid tissue culture

10.1 An Introduction to Brazilian Floriculture

In 2022, Brazilian floriculture generated a monetary value of R$18.4 billion (or approximately US$3 billion) in the domestic market, reflecting a growth rate of 17.0% compared to 2021. This continued the sector's expansion over the past decade. The State of São Paulo accounted for 40% of all sales volume and producers, making it Brazil's leading producer and consumer of flowers (IBRAFLOR, 2024). The Brazilian flower market consists of potted flowers (58% of sales), ornamental plants (24%), and cut flowers (15%) (IBRAFLOR, 2024).

The local flower industry in Brazil devotes most of its work to producing plants purchased from the catalogs of foreign companies, often called "the breeders". These large foreign companies handle sizeable genetic banks that feature species sampled from multiple regions worldwide and use this germplasm for the development, propagation, and copyright registration of new varieties. These corporations also manage the production and sale of cloned plantlets from registered plant cultivars (Cardoso, 2013),

*Corresponding author: jeancardoso@ufscar.br

DOI: 10.1079/9781800627895.0010

primarily obtained through effective *in vitro* micropropagation techniques (Cardoso, 2017; de Oliveira *et al.*, 2019).

However, in recent decades, many of these multinational "breeders" have chosen to lease their licenses to local partner laboratories due to significant fluctuations in foreign exchange rates. These laboratories then perform *in vitro* propagation of copyrighted cultivars, thereby reducing the final production cost of plantlets for export. As Brazil and other developing countries depend on technologies developed abroad, the main consequence is the immense dependence on registered cultivars and licensed technologies developed by breeders. However, local growers and consumers face a foreign currency outflow when buying micropropagated plantlets, with much of the price going to royalties for cultivars from international breeders. This increases production costs, which typically become the primary expense of a flower production system, often exceeding the cost of labor involved in cultivation and production (IBRAFLOR, 2024).

Some of these problems related to the use of foreign varieties by countries lacking these technologies could be tackled by developing local cultivars and technologies adapted to the realities of the Brazilian floricultural market. The development of local technologies and varieties could result in several benefits. These may include better conservation of biodiversity, increased awareness and appreciation of Brazilian plant biodiversity, production of cultivars and varieties better adapted to the Brazilian climate through plant breeding done under the same environmental conditions faced at production greenhouses, strengthening of family farming businesses focused on flower production, a reduction in costs associated with the local supply of plantlets, the establishment of public or private companies involved in plant breeding, the creation of copyrighted flower varieties, and job generation (Cardoso, 2013, 2014; Duval, 2014; Cardoso *et al.*, 2016; Stulzer *et al.*, 2019).

Over the past decade, Brazil and other South American countries have progressed in floriculture, an essential sector of their economies. To address a technological gap in this field, the flower industry and several public universities have initiated advanced breeding programs for various flower species, focusing on plant genera of local origin. These initiatives leverage the wealth of native South American flora and the possible domestication of relatively underexplored species from an ornamental perspective. The activities covered by these programs include the collection and development of germplasm banks, studies in reproductive biology, and hybridization initiatives. Its principal objective is to assess the feasibility of creating commercial hybrids with characteristics that allow them to compete successfully in domestic and international ornamental plant markets.

10.2 The Dominance of Exotic Orchids in the Brazilian Market

Orchids are the most commercialized potted flowers globally, including in Brazil, with the genus *Phalaenopsis* (*Orchidaceae*) being the most prominent, surpassing kalanchoes, calandivas, chrysanthemums, and anthuriums (Chen, 2020; IBRAFLOR, 2021; Yuan *et al.*, 2021). In the Brazilian orchid market, as well as globally, notable ornamental genera include *Phalaenopsis*, *Dendrobium*, *Cymbidium*, and *Vanda* (all of Asian origin), along with *Oncidium* and its hybrids like *Colmanaras* and *Cambrias* (of tropical American origin) (Fig. 10.1).

Of those genera, *Phalaenopsis* and *Dendrobium* stand out as the most intensely bred and selected. Breeders based in the Netherlands and Asian countries market their clonal plantlets, which are then supplied to flower producers in Brazil as cloned plantlets protected or licensed for cultivation in exchange for royalties. These countries have dominated the global market for developing new floral varieties and research and development for large-scale cloning techniques through micropropagation (Cardoso, 2013; Teixeira da Silva *et al.*, 2016; Cardoso *et al.*, 2020).

Phalaenopsis is unquestionably the most traded orchid in the world, including Brazil. For instance, in the Netherlands, 83% of the area cultivated with orchids is dedicated to the genus *Phalaenopsis*, with approximately 137 million pots sold in 2016 alone (Chen, 2020). In Brazil, reliable statistics on this matter are lacking; nonetheless, I estimate that for every 10 orchids sold, seven are *Phalaenopsis*, and that possibly about 10 million pots of *Phalaenopsis* orchids are

Fig. 10.1. Some of main groups of ornamental orchids used in Brazilian floriculture. A– C) *Brassocattleya* orchid hybrids. D) *Oncidium* group. E) *Phalaenopsis* group. F) *Dendrobium* group. Author's own image.

sold annually in the largest flower marketing and distribution center in Latin America, called the Veiling Holambra (https://veiling.com.br/). This leading position can be attributed to the fact that *Phalaenopsis* possesses several desirable features to the flower market, including: (i) a long shelf life in retail settings owing to its resilience to long-distance transportation; (ii) a wide array of flower colors and inflorescence shapes (Yuan *et al.* 2021); (ii) its early and uniform flowering through temperature regulation. This allows for year-round production of flowers in greenhouses equipped with air-conditioning (Cardoso, 2017; Cardoso *et al.*, 2020). Finally, and no less critical, *Phalaenopsis* features a compact plant structure imparted by their showy, succulent green leaves, and slender, long flower stems that carry multiple large and well-rounded flowers. Taken together, the combination of all these traits makes competition from other genera very difficult.

In Brazil, other notable exotic orchids include *Dendrobium*, the 'Nobile' group with numerous short inflorescences along the pseudobulb, and the 'Denphale' group with single or double terminal inflorescences. The genera *Cymbidium* and *Vanda* are also featured in the market, though less prominently than *Phalaenopsis*.

10.3 Orchid Plant Breeding Through Biotechnology

The use of biotechnology in creating and selecting new varieties in *Phalaenopsis* has increased substantially in recent years because of their significant economic relevance and the associated pressure to develop new products. Nevertheless, the number of new hybrids developed with the assistance of biotechnology is still reduced compared to conventional methods (Iiyama *et al.*, 2024). In Brazil, there is neither official information nor estimates regarding the production or sale of transgenic orchids. Still, the market demand for novelty and the need for greater plant tolerance to stress, mainly because of climate change, may push toward adopting biotechnology for quick plant breeding projects. The discovery in Taiwan of a mutation caused by somaclonal variation in plants grown *in vitro* led to the selection and cultivation of "Harlequins," a group of *Phalaenopsis* hybrids. These hybrids feature irregular color patterns on petals and sepals, with spots in red to ruby tones (Fig. 10.1E; Iiyama *et al.*, 2024). This breakthrough revitalized biotechnology's role in orchid breeding programs.

Although *Phalaenopsis* is the current market leader, both worldwide and in Brazil, searching for novelty and innovation is one of the factors that drives and stimulates the floricultural and ornamental market, where this type of consumer is adept at acquiring new products with new characteristics, such as color, shape, and the presence of odors emitted by flowers. In this context, in addition to the demand for products that already exist and feed the market on a large scale, as in the case of the genus *Phalaenopsis*, there is also a search for species and groups of cultivars that bring new interesting characteristics that are different from market standards, in addition to increased resistance to some diseases, as well as the quality and uniformity of flowering (Cardoso, 2017; Li *et al.*, 2021). These new markets aim to supply the flower market more broadly but also attack market niches, such as those related to the collection of rare species and hybrids, as well as specific markets that are more familiar with novelties or groups of species of interest.

It is important to note that in Brazil, the demand for new floriculture products, such as new species and groups of cultivars, is high among end consumers. These end consumers have high expectations for new products sold in retail, especially those related to the flowers' size, shape, and color. The Brazilian orchid market is well supplied by medium to large flowers with intense and vibrant colors, vigorous leaves, and no stains of any kind, requiring horticulturists to use greenhouse cultivation technology intensively. In contrast, though small, the market for local and rare orchid species demands high-value traits aimed at knowledgeable collectors and exhibitors. Supplied mainly by Orchid Associations during local fairs, these species are evaluated and scored by judges based on floral perfection. Typically, these orchids are bred and sold by collectors and small growers locally or online.

The breeding, selection, and trade of cultivars aimed at the general public is very different from that of flower collectors, who are very demanding but pay a high price for a single plant. In the case of collectors, the goal of breeding is associated with what they perceive as the key idiosyncratic traits of a species, however, these criteria may not match the criteria sought after and marketed by large-scale growers, such as resistance to pests and diseases, short cultivation cycles, broad tolerance to temperature and water conditions, and excellent uniformity at flowering.

Regarding orchid flower color, some desired traits in orchids are difficult to achieve using conventional breeding methods and are considered challenging for breeding. An interesting example is the search for atypical colors, such as blue, red, and black, rare in orchids. In the case of blue flowers, a local *Phalaenopsis* breeder has for several years been able to use a synthetic dye applied to the plants at the time of flowering to convert white or pink flowers into blue flowers, meeting market expectations for this type of color. Finding species or hybrids with true blue pigmentation in their flowers is still tricky. The most common products sold naturally in these cases are plants called *Coerulea* varieties, such as those derived from the species *Vanda coerulea* (from India), which have a light blue to purple hue that is highly valued by the collector's market but has a negligible impact on large-scale Brazilian flower production. Similarly, true red is difficult to find in orchids and non-existent in *Phalaenopsis*.

In contrast, color patterns that are remarkably close to red are found naturally in genera that occur in the tropical Americas, such as *Cattleya* and a few intergeneric hybrids. Among the orchids with blackish coloration are some *Oncidium* hybrids such as Sharry Baby (*O.* Jamie Sutton x *O.* Honolulu), in which some of those genotypes show a dark brown tone, close to black, and are commonly sold under the name "chocolate orchid" because of their color, but also because of their sweet and mellow aroma. These *Oncidium* and the more common yellow-flowered ones lead the market in Brazil.

Other challenges to obtaining new orchid cultivars through conventional plant breeding include high heterozygosity and variation, and a long juvenile growth phase, as is the case for most groups of orchids that are either traded or have ornamental potential (Cardoso *et al.*, 2016). These constraints are made worse by the dearth of studies on reproductive biology, limited knowledge on the inheritance of traits of interest, and low efficiency during propagation by tissue culture. The combination of all these technical limitations causes delays in the selection processes. This results in very low success rates, which is evidenced by the low competitiveness of local varieties when placed side by side with foreign orchids. Regarding the long juvenile phase of orchids, which in *Cattleya* can exceed six years from the time of germination, we have achieved significant progress by adopting a systematic selection method that starts during *in vitro* cultivation and continues throughout vegetative development under *in vitro* and greenhouse conditions, in which we select plants that show quick growth up to the bud formation stage. When such traits are selected from the beginning of cultivation, we have observed an increase in the percentage of plants that show very early flowering compared to their parents, with flowering often occurring as early as 18–24 months of cultivation in the greenhouse.

10.4 Main Breeding Methods Used in Brazilian Orchids

Generally, Brazil's main orchid breeding scheme involves crossed fertilization between local species and outstanding hybrids that feature interesting traits, followed by mass selection of progenies under greenhouse conditions (Faria *et al.*, 2015; Stulzer *et al.*, 2019). However, methodological adjustments in the crossing and selection processes have greatly helped to introduce exciting new features, such as quicker flowering and compact size in *Cattleya*, chiefly brought about by: (i) systematic selection of plants with better growth performance; (ii) *in vitro* cultivation of seedlings; (iii) selection and continuous transfer to new medium of progenies with quick growth until plantlets are ready for greenhouse cultivation; and (iv) selection of plantlets for quick growth and robustness until the flowering stage. By resorting to this technique, it has been possible to shorten the flowering time of commercial cultivars to just two years, placing the cultivation time of our *Cattleya* hybrids (Cardoso, 2010; Cardoso *et al.*, 2016) in close distance to genera of greater commercial importance such as *Oncidium* and *Dendrobium* (Cardoso *et al.*, 2016).

It is worth acknowledging that the benchmark criteria for selection across orchid genera, especially when it comes to early flowering, have been set by *Phalaenopsis*, with a cultivation time of only 18 months to flowering, thus *Phalaenopsis* and its cultivation system are considered the most advanced and demanding. In addition, plant breeding in *Phalaenopsis* has expanded into the study of its reproductive biology, especially regarding the reasons for the success of fertilization after pollination. This is because the use of interspecific and intergeneric crosses is widespread in this group of plants and is of particular interest in the development of new hybrid cultivars, where it is possible to observe the pollen viability of the crosses and seeds obtained, in addition to the germination and *in vitro* development of the hybrids obtained from these crosses (Castro and Singer, 2019). From these studies, it would also be possible to systematically evaluate transmitted traits with high heritability to select more interesting parents that transmit traits of interest in multiple crosses to generate hybrids.

10.5 Orchid Breeding in Brazil with Native Species

Brazil has one of the greatest diversities of orchid species worldwide, although the exact number is still controversial. However, it is estimated that there are between 2,300 and 3,500 species, representing 10–15% of all catalogued orchid species worldwide. Many of these species are known as micro-orchids, generally of tiny flowers, with little impact on the conventional flower market. Other groups, although having plants of commercial size and flowers, have the short lifespan of their flowers as their main disadvantage, another trait that also makes the transport and market of these plants difficult.

Some subtribes stand out in this context, such as *Oncidiinae* and *Laeliinae*, which contain an excellent diversity of plants with different shapes, colors, and sizes of inflorescences and flowers that really make them stand out as ornamentals. Considering these two subtribes of orchids, one of the approaches to genetic improvement in orchids has been using native species, which are used in crosses with commercial hybrids or even with other species, in the latter case generating so-called primary hybrids. In the studies carried out by our research group, we have concentrated our efforts on the use of some native species with characteristics of interest to be added to the future generation of commercial hybrids, as in the case of the use of *Brassavola* species in crosses with hybrids and other species in *Laeliinae*, generating the so-called intergeneric hybrids called *Brassocattleya (Fig. 10.1A, B,C)*. Preliminary results obtained by our research group have shown that practically all the hybrids obtained from these crosses have characteristics of interest in the floricultural market, such as ease of cultivation, resistance to pests and diseases, great adaptability to weather extremes (such as heat waves, which have become common in Brazil), compact size, and early flowering, which generally occurs after 12 to 24 months of greenhouse cultivation depending on the genotype. *Brassavola* species are found in the state of São Paulo, the largest flower producer in Brazil, and are responsible for 40% of the total flowers produced in the country. These traits and the fact that they evolved in this habitat and nowhere else allow for the transfer of valuable alleles related to hardiness and quick growth of hybrids, which is quite desirable given the current context of weather extremes experienced during the last 10 years in Brazil. Another feature of some species of *Brassavola* is the occurrence of up to four flowerings per year, in all seasons. Few species in *Laeliinae* have this characteristic, and hybrids of this subtribe conventionally have a single annual flowering, generally very concentrated in specific seasons. Therefore, we aimed to determine whether this trait can be transmitted through multiple crosses with this species.

In *Oncidiinae*, we observed phenomena analogous to those observed in the *Laeliinae*, with species that occur naturally and exhibit a high degree of tolerance to adverse climate conditions. In these species, the flowers' size, shape, and color limit their direct use as ornamentals. However, the results from our group have already demonstrated the possibility of obtaining hybrid plants with other genera of greater ornamental importance, such as *Oncidium* and *Miltonia*, allowing the development of a pipeline of elite hybrids that have benefited from these parental genomes. The objectives were flower color, compact size, and ease of cultivation under the current climatic conditions.

Consequently, and by the experience accumulated to date with orchid breeding, the combination of conventional genetic improvement methods using interspecific and intergeneric crosses between species and biotechnological tools, such as *in vitro* cultivation techniques that significantly increase the efficiency of germination and *in vitro* growth of seedlings from these crosses, has been a powerful tool for breeding programs in orchids and other plant species grown as ornamentals (Teixeira da Silva *et al.* 2016; Cardoso *et al.* 2018). *In vitro* cultivation techniques also allow large-scale *in vitro* cloning of these plants after selecting the most interesting genotypes. This approach expedites the dissemination of new cultivars to producers.

In this context, two techniques should be highlighted that contribute to and accelerate orchid breeding programs, thereby enhancing the profit margins of breeders, both in the development of new hybrids by *in vitro* germination and in the large-scale production of clonal plantlets free from disease and with phenotypic uniformity for the Brazilian flower market: (i) *in vitro* asymbiotic germination; and (ii) cloning using micropropagation techniques.

10.6 Asymbiotic Germination and *In Vitro* Seedling Selection

Although orchids are known for their interaction with mycorrhizal fungi during germination in nature (symbiotic germination), mainly because of the limited reserves during embryo development, the commercial production of ornamental orchid seedlings has been achieved exclusively by asymbiotic germination, that is, without the presence of one or more symbiotic microorganisms. Although symbiotic germination is interesting from the point of view of studies to improve germination and the complex interactions that occur in this plant family, especially in some groups of species in which asymbiotic germination is inefficient, most species and hybrids used as ornamentals do not depend on this type of interaction for germination. A culture medium containing an easily convertible sugar source, such as sucrose, can replace them. This dramatically simplifies the germination processes and makes this technique a protocol in *in vitro* culture laboratories, allowing high rates of germinated seed recovery.

Because one of the characteristics of orchids is the production of hundreds of thousands, and in some cases even millions, of seeds per fruit, the source of genetic material is abundant. However, these numbers must be placed in the context of selection for breeding. Although many seeds are produced in some crosses, only a small proportion may contain viable embryos. This is probably due to the degree of hybridization and capacity for genome recombination between different species, or even in crosses with hybrids or between hybrids. Thus, the abundance of seeds does not always mean an abundance of plants, which requires a high-performance culture medium and a protocol that allows the breeder to recover hundreds, dozens, or even a few hybrid individuals. In this context, the most successful protocol is asymbiotic germination. The controlled cultivation conditions and the reduction of the factors and processes necessary for germination constitute a valuable tool for rapidly recovering and producing seedlings from crosses.

Thus, asymbiotic germination requires several steps to be successful, including pollination (Fig. 10.2A), fruit development (Fig. 10.2B), collection of seed pods (e.g., capsules) at the time of ripening or the beginning of dehiscence, pod removal, pre-drying and appropriate storage of seeds (Fig. 10.2C), asepsis, *in vitro* release of seeds, and germination (Fig. 10.2D) and subculturing of germinated seedlings (Fig. 10.2E) until the time of acclimatization to greenhouse conditions (Fig. 10.2F,G). Our research group's protocols include harvesting the seed pods at the onset of yellowing and dehiscence, forcibly opening the pods, and collecting the seeds in a clean Petri dish. After collection, the seeds were kept at room temperature (25–30°C) for up to three days before being placed in plastic tubes and stored at refrigeration temperature (5–8°C). After collection, the purpose was to allow the seeds to lose excess water, thereby increasing the storage time at low temperatures. In this way, seeds can be stored for up to 24 months after collection without significant loss of germination. However, it should be noted that crosses where we obtain very few seeds with embryos are placed for asymbiotic germination immediately after collection.

Seed asepsis is very simple to achieve by using a 0.2–0.3% sodium hypochlorite solution for 10–20 minutes with constant agitation. Subsequently, the solution containing the seeds was left to stand for another two minutes. This pro-

Fig. 10.2. Orchid conventional breeding system using asymbiotic germination to accelerate and improve progenies selection. A) controlled pollination (pollinia in stigma) aimed at hybridization of orchids. B) Fruit development after pollination. C) Fruit dehiscence and seed harvest. D) Efficient asymbiotic germination and development of seedlings under *in vitro* conditions. E) *In vitro* elongation and rooting of seedlings. F) Acclimatization of *in vitro*-derived seedlings. G) Successful plantlet acclimatization in substrate with proper root and shoot development. H) Adult plant selected from hybrid progeny as new potential cultivar for orchid flower market. Author's own image.

cess separates most seeds containing embryos (denser than the solution) from those remaining as supernatants (lighter than the solution), allowing for the initial separation of seeds to increase the chances of success during *in vitro* germination. After this procedure, the seeds were washed thrice with autoclaved deionized water. The entire process can be performed in 15–20 ml plastic bottles using a small volume of solution and water for asepsis.

The seeds are then transferred to germinate in Murashige and Skoog (MS) medium (Murashige and Skoog, 1962) modified by halving the concentrations of the macronutrients to just 20–25 g l^{-1} sucrose and 0.1 g l^{-1} inositol and solidified with agar. An essential step in preparing the medium is maintaining the pH at approximately 5.5 instead of the conventionally used 5.8. In our experience, this pH increases the percentage of seeds germinated *in vitro.*

The initial subculture phase involves selecting plants that exhibit accelerated growth, characterized by the development of pseudobulbs, leaves, and roots (Fig. 10.2D), as opposed to those that remain in the form of protocorms or exhibit limited development in certain parts. During this phase, the most developed plants continue to be subcultured until they achieve the seedling standard necessary for acclimatization. This standard, which may vary depending on the specific cross, typically requires seedlings to possess two to three leaves measuring 3–10 cm long and at least two elongated roots (Fig. 10.2E).

Plants reaching this standard were then removed from the culture flasks, washed with running water to remove the culture medium from the roots, and acclimatized in plastic trays containing a substrate based on peat, charred rice husk, and vermiculite with good porosity and water retention (Fig. 10.2F).

10.7 Protected Cultivation of Plants From *In Vitro* Selection

Breeding through the selection process of superior phenotypes continues after acclimatization, with the seedlings remaining in the acclimatization tray for 120–150 days (Fig. 10.2G). During this phase, approximately 100 plants from each cross are acclimatized to greenhouse conditions with temperature control (18–30°C) obtained by the pad-and-fan system, which cools automatically by injecting humidity into the air through cellulose hives. The system is complemented by exhaust fans that remove hot air from the greenhouse and replace it with cold, humid air from the hives. The greenhouse is closed on all sides with agricultural plastic diffusers, and the upper side is complemented by an aluminum shading screen that blocks 70–80% of sunlight in tropical conditions.

After the cultivation period under these conditions, the acclimatized seedlings can be transferred to the cultivation greenhouse, which may or may not be air-conditioned, depending on the orchid species being bred and the local weather conditions. Generally, the genus *Phalaenopsis* and some intergeneric *Oncidium* hybrids require air-conditioned greenhouses for cultivation in Brazil. In contrast, intrageneric *Oncidium* and *Dendrobium* hybrids and intergeneric *Cattleya* hybrids can be cultivated without air conditioning. In this context, the breeder needs to adjust the cultivation settings similar to those used by the grower who will grow the hybrid for selection, because the performance of hybrids depends significantly on the weather to which the progeny is exposed to.

After acclimatization, the best plants from the tray, with the best development of shoots and roots and without symptoms of pests and diseases, are transferred to plastic pots of 8 cm diameter, where they remain under these conditions for 6 to 12 months. Finally, the plants are transferred to final pots, which can be 9, 12, or 15 cm in diameter, depending on the expected size of the hybrid plants. In the case of the *Phalaenopsis* genus, elite plants are grown in pots #12 in Brazil, which are preferred for selection in breeding and cultivation. Our research group has developed compact *Cattleya* plants adapted to pots #9 or 12, similar to what occurs with *Phalaenopsis*. This is a consumer trend and is also aimed at increasing the number of pots per square meter in greenhouses, reducing costs, and meeting the end customer's demands.

Final selection occurs when the plants flower (Fig. 10.2H). In this phase, the flowering characteristics are specific and depend significantly on the objective of the breeder, which may be the selection of plants with early flowering and late flower senescence, both of which are of interest in *Cattleya* hybrids, since most *Cattleya* have a long juvenile cycle (3–7 years of cultivation) and short flower duration after anthesis (maximum 20 days). These characteristics greatly limit Cattleya's large-scale cultivation and commercialization, as they increase production costs and limit commercialization over long distances from the production site. Therefore, we have considered it a priority to develop new cultivars that flower early and have a more extended flowering period, generally 30 days or more. Another highly relevant characteristic of *Cattleya* is the low uniformity of flowering, even in clones of selected hybrids, which makes it challenging to scale up production or produce for specific dates. However, this last characteristic can only be chosen after cloning the previously selected hybrids, which allows the evaluation of several plants. However, it is possible to obtain an idea based on the crosses made. We have identified parent plants capable of transmitting these characteristics to their progeny with good heritability, and it is possible to observe concentrated flowering of the progeny at one time, which shows that even if there is genetic variability in the selection material, it is uniform in terms of flowering time. Thus, it is expected that such material, when cloned, will have a very high uniformity in flowering plants, allowing the commercialization of entire batches on scheduled dates.

In *Phalaenopsis*, on the other hand, in addition to compact size, goals such as plants with two inflorescences and novel colors and shapes are primary objectives for the plant to enter the large market. In *Oncidium* hybrids, compact size, multiple inflorescences, and inflorescences with many colored flowers are the primary breeding goals.

10.8 *In Vitro* Cloning Using Micropropagation

Finally, after selecting plants with superior characteristics and candidates for new cultivars, another biotechnological method is highly relevant for breeders, allowing the plant chosen to be commercialized as a plantlet on a large scale. The biotechnological process that enables the generation of thousands of plantlets from small tissue samples from a donor plant, usually the meristematic regions, is micropropagation. This is because in the various groups of orchids used as ornamentals, the conventional production of plantlets is quite limited by the low rate of the annual output of new shoots and by the limited and slow growth of the plants, which reduces the number of vegetative propagation techniques available to generate plantlets from a plant candidate for a new cultivar. Seeds are abundant, but high species/hybrid heterozygosity prevents maintaining selected traits in offspring. Thus, the only viable method for large-scale plantlet production is using plant tissue culture or the *in vitro* cultivation of plants by cloning superior plants using somatic tissues from the donor.

Micropropagation represents the most widely utilized technique for *in vitro* cloning and large-scale production of cultivar plantlets. In the context of orchids, micropropagation is conventionally executed using a solidified agar culture medium cultivated in growth rooms. This technique, classified as vegetative propagation, has successfully produced thousands of plantlets from minute segments of somatic tissues, predominantly those derived from meristematic regions or tissues containing meristem. In orchids, the micropropagation process is subdivided into the following phases: (i) the mother plant care and cultivation phase, in which both the selection and the necessary and differentiated care of the mother plants occur, aiming at successful *in vitro* establishment; (ii) the *in vitro* establishment phase, which involves the collection of plant tissues to be used and grown in the laboratory, the asepsis process to eliminate potentially contaminating microorganisms from the culture medium, and the initial *in vitro* regeneration process; (iii) the multiplication or proliferation of shoots; (iv) rooting and shoot elongation; and finally, (v) the acclimatization (e.g., hardening) of the plantlets obtained *in vitro* to greenhouse conditions.

10.9 Care and Cultivation of Mother Plants

Caring and cultivation for mother plants is a multifactorial process, but it should also be part of the planning for *in vitro* plant production and cloning. One of the key factors for success in this phase is the correct selection and identification of genotypes to be micropropagated. The selection of the genotypes to be cloned is a key responsibility by the breeder, who must select the materials according to the demand by top customers and the suggestion of new products by the flower market itself. The organization of the set of mother plants (the breeding stock), especially about the correct and precise identification of the cultivars to be propagated, is necessary, allowing the collection of tissue for cloning to be easily identified, and avoiding confusion about the cultivar to be propagated. It is important to emphasize that, as a cloning process, at the end of the micropropagation process, thousands of genetically identical plantlets that are true-to-type are expected to be obtained. Incorrect identification or even the incorrect collection of genetic material that is not identified may result in the marketing of a genotype that is entirely different from the one requested or marketed, which may compromise the reliability of the laboratory.

In addition to genetics, another important aspect at this stage is the phytosanitary issue: the availability of presence of tissues free from disease. Among the main diseases affecting orchids are viruses, including Cymbidium Mosaic Virus (CyMV) and Odontoglossum Ringspot Virus (ORSV), as well as fungal diseases caused by fungi belonging to the genera *Fusarium*, *Alternaria*, *Cercospora*, *Botrytis*, and *Colletotrichum*. Some bacteria also threaten the cultivation of *Phalaenopsis* in greenhouse settings.

Knowing the symptoms and applying preventative products to control these microorganisms could be an interesting strategy to avoid surprises in the stock of mother plants. Systemic fungicides can be used periodically to keep plants free of these pathogens.

In the context of pathogen control, innovations have emerged to increase the sustainability of production systems, including the biological control of pests and diseases using microorganisms. Experiences with biological control using commercially available microorganisms such as *Trichoderma*, *Bacillus*, *Beauveria*, and others in organs used for tissue collection for *in vitro* propagation have not been positive because these microorganisms can become contaminants within tissue culture laboratories. It is important to emphasize that these adverse effects are not associated with plant production and cultivation, where these microorganisms have been beneficial and have helped control pests and diseases in agriculture. However, since these microorganisms are produced in culture media, have high adaptability to these cultivation conditions, and have high proliferation, the risks associated with the introduction of contamination from these microorganisms into *in vitro* plant culture laboratories are high. Recently, we observed large-scale contamination of *in vitro* cultures, resulting in losses of more than 40% of all subcultured material, caused precisely by *Bacillus subtilis*. The source of contamination was precisely in the potting medium where the grower used this microorganism to improve plant rooting and reduce the use of pesticides in cultivation. During *in vitro* culture, microorganisms such as *B. subtilis* compete with the plants for resources and, in most cases, even produce toxins that lead to the death of *in vitro* plantlets.

In the case of viruses that are difficult to control using chemical products, it is vital to keep the plants and the potting mixture virus-free. Reports of transmission of these viruses in orchids have involved hand contact and dirty cutting instruments; therefore, proper cleanliness of materials used for tissue collection, or even the use of disposable blades for this purpose, is a strategy for controlling viruses in orchids. In the case of plants contaminated with viruses, where it is necessary to obtain virus-free plants, combined techniques such as thermotherapy and chemotherapy, together with the cultivation of stem tips, may be required to eliminate the virus from the propagated tissues. It is important to emphasize here that, although the tissues undergo the asepsis process at the time of *in vitro* establishment, the products used at this stage are not systemic and provide only superficial disinfection to prevent contamination of the culture medium with microorganisms that survive on the surface of the plant tissues, whether pathogenic or not.

As in the case of plant pathogens, we recommend that nutrition for mother plants should rely exclusively on chemical fertilizers, avoiding organic fertilizers and those rich in microorganisms. The aim here, especially in the case of mother plants, is to reduce the number of microorganisms to enhance the efficiency of the next phase, which is the *in vitro* establishment of plants and their regeneration, without contaminating microorganisms in the culture medium.

10.10 *In Vitro* Establishment

In vitro establishment is the first laboratory phase of micropropagation and the phase of *in vitro* cultivation. In this phase, the somatic tissues collected in the previous phase undergo processes such as explant selection, asepsis, inoculation, and *in vitro* regeneration.

In the case of orchids, two types of explants have been most commonly used: young shoots in the case of *Cattleya, Oncidium, Dendrobium,* and their hybrids, and flower stem internode segments in the case of *Phalaenopsis* and *Oncidium*. Flower stem segments are pretty interesting, especially in the case of *Phalaenopsis*, a monopodial orchid, where the extraction of basal stem meristems sacrifices the individual to be cloned, since the small pseudobulb, associated with the meristems and placed at the base of the leaves, makes it difficult to extract clean stem tissue without killing the plant. *Phalaenopsis* flower stems are divided into nodes and internodes, with flower buds developing in the upper half or third of the inflorescence. Nodes containing dormant buds remain in the lower part of the inflorescences. These regions include parts of the internodes and the node, which features a dormant bud used for asepsis and *in vitro* culture. In the case of sympodial plants, young shoots from the base of the plants are cut and removed, usually when they are 5–10 cm long and the leaves have not yet begun to open. The shoots are subjected to asepsis, and the tissues surrounding the leaves are removed from the shoots to expose the axillary and apical buds. The buds are then placed into the culture medium for regeneration. The asepsis of these tissues is typically carried out using 70% alcohol for 1 to 5 minutes and a dilution of bleach and water, resulting in a concentration of 1.0–1.5% active chlorine, depending on the species or hybrid, or even the cultivation conditions to which the plants are being subjected.

Finally, after placing explants into the culture medium, these tissues are expected not to die and to be free of contamination by microorganisms. Subsequently, the process of *in vitro* regeneration begins.

The *in vitro* regeneration of orchids for micropropagation occurs mainly in two ways: through adventitious shoots (direct organogenesis) and the so-called protocorm-like bodies (PLBs), the latter being structures similar to zygotic embryos in orchids (protocorms) but of clonal origin, so it is often referred to as somatic embryogenesis. However, a mixture of these two regenerative processes can also be frequently observed in the initiation/establishment culture medium, demonstrating that both regeneration pathways can originate from the same cultured tissue. Preferably, especially in some cultivars that are more sensitive to somaclonal variation, direct regeneration via adventitious shoots should be chosen because regeneration via PLBs can be a source of somaclonal variation, that is, the generation of plants with characteristics different from those of the original plant, which generally appear in flowers and inflorescences (Cardoso *et al.*, 2020).

The culture medium used in this phase may vary according to the group of species and hybrids of interest, but usually MS medium with half the concentration of macronutrients and plant regulators such as 6-benzylaminopurine (0.5 to 4.0 mg l^{-1}) and naphthaleneacetic acid (0.1 to 1.5 mg l^{-1}), results in reasonable regeneration rates. In general, when choosing flower stems, the concentration of plant regulators should be higher. Lower concentrations are necessary to achieve good results when choosing axillary and apical buds.

10.10.1 *In vitro* multiplication or proliferation of shoots

The goal of this phase is to exponentially increase the number of shoots from these selected plants in the least amount of space and in the shortest amount of time. Propagation involves transforming a small amount of tissue from the mother plant or a new cultivar, thus generating thousands of shoots/plantlets through successive subcultures. With each subculture or transplantation of shoots of plantlets, an increase in the number of plantlets is expected, and the number of shoots obtained at the end of cultivation, divided by the number of shoots cultured, gives the multiplication rate. The higher the multiplication rate, the faster the desired number of shoots/plantlets can be retrieved and sold, as more plants can be subcultured and produced in the same amount of time, space, and culture medium. In the case of orchids, a reasonable multiplication rate is between 3.5 and 7.0 shoots/cultured shoot every 60–90 days of cultivation. The MS culture medium with half the concentration of macronutrients, and the cytokinins 6-BAP or 2-isopentenyladenine (2-ip), at concentrations between 0.5 and 4.0 mg l^{-1}, has been widely used for a large number of genera and cultivars, with good results in terms of multiplication.

A critical point in this phase is the number of subcultures in the multiplication phase, with at most four to six subcultures recommended. This restriction is suggested due to the possibility of undesirable mutations occurring in the explants, which could result in a large number of plantlets becoming somaclonal mutants, that is, those with genetic or epigenetic alterations that could modify the visual character of the micropropagated hybrid.

Therefore, realistic production schedules and reasonable sales targets for clonal plantlets in orchid micropropagation should be estimated based on the number of initial available shoots to be cultured multiplied by the standard proliferation rate of that genotype, as well as possible losses that might be incurred during the process. Thus, the number of plantlets to be produced should be properly estimated, and the best time to deliver them could be scheduled well into the future.

10.10.2 Rooting and elongation of shoots

Due to the use of cytokinins, such as 6-BAP, it is common for orchid shoots from the multiplication or proliferation phase to be rootless and short in length. Therefore, after subculturing in the multiplication medium and obtaining enough shoots for cultivation, these shoots must go through an elongation and rooting phase, in which the aim is to produce adventitious roots and increase the size of the shoots, as well as a better differentiation of organs, with the formation and proper definition of pseudobulbs and leaves, under *in vitro* conditions. These characteristics are necessary for *in vitro* shoots to have a better chance of survival in the acclimatization phase when the *in vitro* plantlets face the harsh conditions of a greenhouse.

At this stage, it is therefore expected that the shoots will differentiate into pseudobulbs, leaves, and roots, retaining at least two roots of at least 2 cm in length and three to four well-developed leaves, giving a total aerial part of 4 to 10 cm in length. Logically, depending on the genotype, some of which are compact, these values can be reanalyzed and implemented according to the laboratory's experience.

A critical aspect of this rooting phase is that it requires a lot of space inside the laboratory, representing up to 70% of the laboratory's occupancy rate. This is because the number of plantlets per flask is reduced, allowing each plantlet to develop and reach the appropriate size. For orchids, this phase requires 60 to 90 days of cultivation, and in most cases, at least one subculture is necessary, which would total up to 180 days of cultivation for the elongation and rooting of the shoots. In addition, cultivation under *in vitro* conditions results in significant morphological and physiological changes in plantlets, mainly because of sucrose in the culture medium and cultivation in closed flasks without any gas exchange. This can make acclimatization difficult, even if the plantlets reach sufficient size and differentiation, as these plantlets have difficulty responding to the changes that occur from the shift from *in vitro* to *ex vitro* cultivation.

One solution would be to transfer this phase to a system that better prepares the plantlets for the next acclimatization phase. Systems with increased gas exchange, whether with forced ventilation or even plastic covers containing gas-permeable membranes, can provide benefits for this and the next phase, such as increased photosynthetic rates in *in vitro* plants and lower relative humidity, resulting in better stomatal control ability of the plantlets. In addition, using environments such as greenhouses, which could expose plantlets to shaded sunlight, until 20–30% of full-sunlight intensity, and provide an environment with narrow temperature fluctuations, may result in better hardening of the plantlets, aiming at their successful acclimatization. Logically, this leads to exceptional care, such as using a greenhouse with climate control, at least for temperature and light, and ensuring good cleanliness at the location, avoiding excessive contamination with plant residues and old potting medium. In addition to these advantages, the laboratory would have a greater capacity to multiply different materials, reducing the space occupied by plantlets in the elongation and rooting phase.

10.10.3 Acclimatization phase

The acclimatization phase is one of the most critical factors in the *in vitro* cultivation of orchids, as losses here represent the loss of all resources invested in the plantlets, with a significant increase in production costs. Therefore, the preparation of *in vitro* plantlets for this phase is necessary to improve survival rates. Some examples of methods that can be used in the previous phase to improve acclimatization help to prepare plantlets for this phase better. However, inherent acclimation concerns must be considered, including light, temperature, relative humidity, irrigation, nutrition, substrate, and plant health.

Luminosity between 10 and 20 thousand Lux (200–400 μmol s^{-1} m^{-2}), temperatures from 18 to 30°C, and high relative humidity are the minimum conditions for good acclimatization; therefore, a greenhouse with controlled luminosity and temperature is necessary for acclimatization of plantlets throughout the growing year. These conditions may allow the successful readaptation of *in vitro* plants kept in a laboratory environment at mild temperatures, low light levels, and low gas exchange. A pad-and-fan system or misting with foggers may provide reasonable greenhouse climate control and maintain high relative humidity in the environment to prevent excessive dehydration of the plantlets.

Substrates based on peat, carbonized rice husks, and vermiculite have shown promising results in acclimating orchid plantlets from *in vitro* culture. Nevertheless, other materials such as coconut fibers can also be used successfully. A new product that we have already tested and that has worked very well with *Phalaenopsis* is plugs or paper pots, a product with a compacted substrate that facilitates growing plantlets. One of the most critical conditions for the substrate in the acclimatization of orchids is the transmission of pathogens, and high-quality substrate is required for use in this phase. It is important to note that plantlets from *in vitro* cultivation are grown for 12–18 months in an environment without microorganisms, with a deficient defense system against pathogens.

Typically, flasks are opened one day prior to the removal of plantlets for acclimatization. The plantlets can then be removed from the flasks and washed in good-quality running water to remove any residues of the culture medium attached to the roots. Phytosanitary treatments with protective fungicides can be applied, and plantlets can be planted individually in trays or groups of plantlets.

Watering should be more frequent during the first 4 weeks of cultivation, keeping the substrate well-moistened, without dry periods. After these 4 weeks, the watering frequency can be reduced, providing moist and slightly moist substrate periods to bring the plant closer to the growing conditions.

The nutrition must be balanced, containing all macro- and micronutrients, applied at least once a week by fertigation, with an electrical conductivity of 0.6 to 0.7 mS cm^{-1}. In this phase, urea can be used as a nitrogen source, but care must be taken in its management because of possible salinization and acidification of the substrate, which can lead to plantlet death.

Pest and pathogen control should be based on the growing region, available and recommended pesticides for the crop, and the target pathogens.

10.11 Conclusions

In Brazil and other parts of the world, orchids constitute a significant segment of the flower and ornamental markets, with the genus *Phalaenopsis* being particularly predominant. Other notable ornamental genera, including *Oncidium*, *Dendrobium*, *Cattleya*, and *Cymbidium*, and their multigeneric hybrids, also play a crucial role in Brazilian floriculture. Despite this, Brazil remains heavily reliant on technologies developed internationally, particularly for developing new cultivars and producing high-quality plantlets through biotechnological methods. Some national initiatives to breed orchids and other floral species have emerged, primarily within public and private institutions. Although cultivars are being introduced for the public sector, there is a need for stronger collaboration between breeding researchers and the private sector to ensure that introduced cultivars meet the minimum market requirements. In response, private initiatives have arisen, with companies and flower producers transitioning into breeding enterprises, proposing the introduction of new cultivars within a production context under evaluation. Biotechnology thus contributes to the improvement of orchids through asymbiotic seeding, which facilitates *in vitro* germination of progeny from breeder-made crosses, and by cloning of selected genetic material via micropropagation. The application of additional biotechnological tools, such as transgenic and genome editing techniques, remains contentious, with well-developed methodologies and regulatory controls that require substantial financial investment by companies for product commercialization. These factors have not significantly encouraged the development of differentiated cultivars using these technologies for ornamental plants in Brazil. Consequently, conventional selection through crossbreeding, combined with biotechnology to expedite traditional programs and enable large-scale propagation and production of high-quality plantlets, will remain vital for orchid production in Brazil for the foreseeable future.

Acknowledgements

JCC thanks the São Paulo Research Foundation (FAPESP) grant n° 2018/20673-3, to Conselho Nacional de Desenvolvimento Científico e Tecnológico (National Council on Science and Technology Development, CNPq, Brazil) for funding projects 311083/2018-8 and 317162/2021-7. This study was financed in part by the Coordenação de Aperfeiçoamento de Pessoal de Nível Superior - Brasil (Directorate for the training of university personnel, CAPES) - Finance Code 001.

References

Cardoso, J.C. (2010) *Laeliocattleya* 'Brazilian Girl Rosa': cultivar de orquídea para cultivo em vaso. *Horticultura Brasileira* 28, 378–381.

Cardoso, J.C. (2013) Melhoramento de espécies ornamentais como estratégia para o desenvolvimento e autossuficiência do setor. *Horticultura Brasileira* 31, 171.

Cardoso, J.C. (2014) Levantamento de espécies da família Orchidaceae em Botucatu: potencial ornamental e cultivo. *Horticultura Brasileira* 32, 7–13.

Cardoso, J.C. (2017) *Ionocidium* 'Cerrado 101': intergeneric hybrid with high quality of blooming. *Ornamental Horticulture* 23, 351–356.

Cardoso, J.C., Martinelli, A.P. and Teixeira Da Silva, J.A. (2016) A novel approach for the selection of *Cattleya* hybrids for precocious and season-independent flowering. *Euphytica* 210, 143–150.

Cardoso, J.C., Gerald, L.T. and Teixeira Da Silva, J.A. (2018) Micropropagation in the twenty-first century. In: Loyola-Vargas, V., Ochoa-Alejo, N. (eds) *Plant cell culture protocols, methods in molecular biology, vol 1815*. Humana Press, New York, USA. DOI: 10.1007/978-1-4939-8594-4_2

Cardoso, J.C., Zanello, C.A., and Chen, J.T. (2020) An overview of orchid protocorm-like bodies: mass propagation, biotechnology, molecular aspects, and breeding. *International Journal of Molecular Sciences* 21, 985. DOI: 10.3390/ijms21030985

Castro, J.B., Singer, R.B. (2019) A literature review of the pollination strategies and breeding systems in Oncidiinae orchids. *Acta Botanica Brasilica* 3. DOI: 10.1590/0102-33062019abb0111

Chen, C. (2020) The fundamental issue in the *Phalaenopsis* industry in the Netherlands. *Journal of Agriculture and Forestry* (農林學報) 67, 89–100. DOI: 10.30089/JAF.202006_67(2).0002

Duval, C.M. (2014) A produção de flores e a agricultura familiar. *Horticultura Brasileira* 32, 241.

Faria, R.T.; Colombo, R.C. and Hoshino, R.T. (2015) *Oncidium sarcodes* x *Oncidium* Aloha 'Iwanaga': new option of *Oncidium* hybrid. *Horticultura Brasileira* 33, 131–134

IBRAFLOR (Instituto Brasileiro de Floricultura/Brazilian Flower Cultivation Institute) (2021) O mercado de flores no Brasil, 2021. Available at: https://354d6537-ca5e-4df4-8c1b-3fa4f2dbe678.filesusr.com/ugd/b3d028_e002f96eeb81495ea3e08362b49881a3.pdf (accessed 8 September 2021)

IBRAFLOR (Instituto Brasileiro de Floricultura/Brazilian Flower Cultivation Institute) (2024) Flores em números, mercado interno PIB. Available at: https://www.ibraflor.com.br/n%C3%BAmeros-do-setor-c%C3%B3pia, (accessed 28 March 2025)

Iiyama, C. M., Vilcherrez-Atoche, J.A., Germanà, M.A., Vendrame, W.A. and Cardoso, J.C. (2024) Breeding of ornamental orchids with focus on *Phalaenopsis*: current approaches, tools, and challenges for this century. *Heredity* 132, 163–178. https://doi.org/10.1038/s41437-024-00671-8

Li, C., Zhao, Y.D., Wu, S., Liu, Z., Zhai, J.A. (2021) Review for the breeding of orchids: current achievements and prospects. *Horticultural Plant Journal* 7, 380–392. DOI: 10.1016/j.hpj.2021.02.006

Oliveira, B.C., Oliveira, M.E.B.S. and Cardoso, J.C. (2019) Feasibility of the new method for orchid *in vitro* rooting using liquid and chemical sterilized culture medium under different sucrose concentration. *Ornamental Horticulture* 25, 263–269.

Stulzer, G.C.G., Hoshino, R.T., Suzuki, A.B.P., Paula, J.C.B., Alves, G.A.C. and Faria, R.T. (2019) Primary hybrid of *Cattleya forbesii* x *Cattleya loddigesii*, a new Brazilian orchid. *Crop Breeding and Applied Biotechnology* 19, 364–367. DOI: 10.1590/1984-70332019v19n3c50

Teixeira Da Silva, J.A., Jin, X., Dobránszki, J., Lu, J., Wang, H., Zotz, G., Cardoso, J.C. and Zeng, S. (2016) Advances in *Dendrobium* molecular research: applications in genetic variation, identification and breeding. *Molecular Phylogenetics and Evolution* 95, 196–216. DOI: 10.1016/j.ympev.2015.10.012

Yuan, S.C., Lekawatana, S., Amore, T.D., Chen, F.C., Chin, S.W., Vega, D.M. and Wang, Y.T. (2021) The global orchid market. In: Chen, F.C. and Chin, S.W. (eds) *The orchid genome. compendium of plant genomes.* Springer, Cham. DOI: 10.1007/978-3-030-66826-6_1

11 Genetics and Propagation of *Sobralia* Orchids

Pablo Bolaños-Villegas*

Fabio Baudrit Agricultural Research Station, University of Costa Rica, La Garita District, Alajuela, Costa Rica; Lankester Botanical Gardens, University of Costa Rica, Dulce Nombre District, Cartago, Costa Rica; Department of Food and Agricultural Sciences, Rodrigo Facio Campus, School of Agronomy, University of Costa Rica, Montes de Oca County, San Jose, Costa Rica

Abstract

Sobralia is a genus of terrestrial orchids native to the tropical regions of the Americas, distinguished by the fleeting yet striking beauty of its large and symmetrical flowers. Limited efforts have been made in plant breeding within this genus, resulting in most specimens being sourced from wild populations. Research conducted at Lankester Gardens indicates that *in vitro* seed germination rates are high, suggesting the potential for hybridizing and propagating species with advantageous traits such as extended flower longevity and fragrance. Future research may concentrate on the greenhouse acclimatization of tissue culture plants and polyploidization, aiming to establish a robust pipeline of hybrids exhibiting vibrant flower colors.

Keywords: *Sobralia*, terrestrial orchids, tissue culture, cytogenetics, flower withering.

11.1 Introduction

Sobralia is a diverse genus of neotropical orchids. About 200 species are distributed across the Americas except for the West Indies (Dressler *et al.*, 2016). Most species exhibit gregarious flowering with large symmetrical, yet ephemeral flowers (Dressler *et al.*, 2016). In Costa Rica, illegal vendors commonly sell these orchids on the street. Poaching and illegal trade may lead to the extinction of wild orchid species; hence, research into the genetics and propagation of these beautiful plants will contribute to their preservation.

Sobralias are difficult to classify, and hybrid swarms involving at least three species are common in Cartago, Costa Rica (Dressler and Pupulin, 2014). These hybrid swarms lose the ability for synchronized flowering (Dressler and Pupulin, 2014). However, for any prospective plant breeder or biologist, information is lacking on the shape and number of chromosomes that would allow for identifying putative hybridization and speciation events, as has been done in the orchid genus *Phalaenopsis* (Bolaños *et al.*, 2008). In *Phalaenopsis*, odd numbers, irregularity in size, and defects in segregation during meiosis suggest combinations of chromosomes

*Corresponding author: pablo.bolanosvillegas@ucr.ac.cr

(Eds P. Bolaños-Villegas and F-C. Chen)
DOI: 10.1079/9781800627895.0011

from different genomes. As in *Phalaenopsis*, in *Sobralia*, the base number of chromosomes was estimated with the cultivation of more than 40 species at the Lankester Gardens at the University of Costa Rica (Dressler *et al.*, 2016).

In this project at the gardens, microscopic observation of meiocytes harvested from *Sobralia* flowers did not reveal any obvious defects in the mating or segregation of sexual chromosomes (meiotic) (Fig. 11.1), and normal tetrad formation was efficient, always more than 90% (Table 11.1), unlike commercial orchid hybrids that are not fertile (Arida *et al.*, 2021; Tanaka and Kamemoto, 1960). In addition, no species exhibited morphological differences within the chromosome complement, except for *S. fenzliana*, which showed distinct elongated chromosomes (Mora *et al.*, 2018). Notably, the number of chromosomes varied across species, with the diploid complement (2n) ranging from 16, 24, 28, 32, 34 to 48, which suggests a yet-to-be-identified speciation mechanism operating across these karyotypes (Mora *et al.*, 2018). Initially, holocentricity was suggested to be involved in both chromosome number evolution and highly efficient chromosome segregation. However, immunostaining experiments with commercial antibodies against the phosphorylation of threonine-120 of histone H2A (H2AT120ph) (Oliveira *et al.*, 2024) suggested that kinetochores are discrete (e.g., monocentric) in *S. danjanzenni*, *S. chrysostoma*, *S. kerryae*, *S. helleri*, *S. fenzliana*, and *S. virginalis* (data not shown).

The paradoxical finding of the normal formation of meiotic tetrads despite a significant divergence in chromosome number (Table 11.1) may also suggest a recombination mechanism responsible for enforcing the mating of meiotic chromosomes in *Sobralia*. The mechanism awaits identification, but several processes are known to regulate base recombination levels across eukaryotic genomes; such mechanisms include SUMOylation (Bhagwat *et al.*, 2021) and timely resolution of recombination

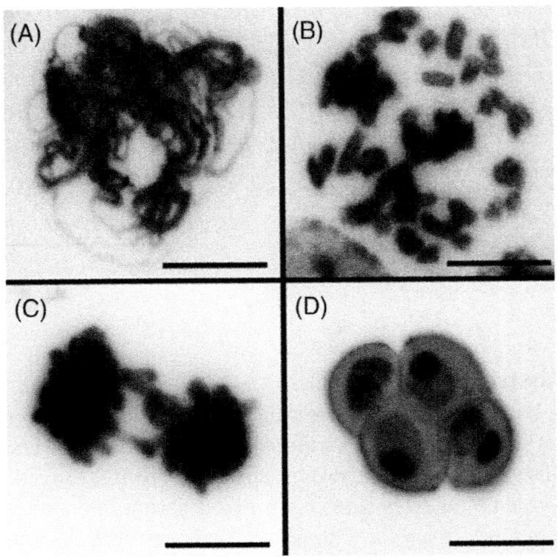

Fig. 11.1. Meiosis in *Sobralia helleri* (2n = 24 ± 1). A) Pachytene stage. B) Possible somatic metaphase. C) Telophase I. D) Tetrad stage. Chromosomes were stained with 4′,6-diamidino-2-phenylindole (DAPI). Scale bar: A and B, 20 μm; C and D 10 μm. Fixation involved propionic acid solution and pretreatment with 8-hydroxiquinoline solution for enhanced condensation. Author's own image.

Table 11.1. Production of regular tetrads following meiosis and prospective chromosome numbers at meiotic metaphase I in species of *Sobralia*. Author's own table.

Species	Tetrads (%)*	Chromosomes (2n)**
S. amparoe	89.3 ± 8.5	30 ± 1
S. atropubescens	80.0 ± 6.6	–
S. bouchei	88.4 ± 5.3	34 ± 2
S. bradeorum	89.7 ± 1.8	48 ± 0
S. carazoi	85.7 ± 5.1	–
S. crispissima	83.1 ± 10.2	16 ± 1
S. danjanzenii	93.5 ± 0.7	32 ± 2
S. fenzliana	87.5 ± 1.0	28 ± 1
S. geminata	89.5 ± 1.0	32 ± 2
S. helleri	86.0 ± 2.1	24 ± 1
S. rosea	93.1 ± 2.1	24 ± 1

Data are mean ± SD and were estimated from three biological replicates. *n = 100. **n = 20.

intermediates by the RecQ helicase (Whitbread *et al.*, 2021). Moreover, because a base chromosome number remains to be determined, any presumed hybridization events in presumptive hybrid swarms will require resolution by molecular characterization instead of microscopy-based cytogenetics.

In parallel, *in vitro* pollen germination was found to be normal in the species (Mora *et al.*, 2020). At this stage of the project, the COVID-19 pandemic began, and collection and analysis activities were reduced to a minimum. However, this time was used to test the species' ability to pollinate themselves and others. This aspect had not been studied and there were no data available, except for the occasional presence of capsules resulting from random pollination events, presumably due to the activity of bees or hummingbirds (hummingbird nests are often found on these plants).

11.2 Self- and Cross-Pollination Dynamics in *Sobralia*

From February 2020 to September 2023, more than 100 flowers of Sobralia species were crossed, from the species *S. atropubescens, S. bouchei, S. crispissima, S. chrysostoma, S. danjanzenii, S. decora, S. dorbiguyana, S. fenzliana, Sobralia x gloriana, S. gloriana, S. helleri, S. kerryae, S. leucoxantha, S. luteola, S. macrantha, S. nutans, S. rogersiana* and *S. turrialbina* (Fig. 11.2). These plants are grown in dedicated greenhouses under a fertigation program, with the plant substrate changed at least once a year (a mixture of charcoal, wood bark and slow-release fertilizer). Fungicides and bactericides are also sprayed periodically to keep the plants healthy.

The crossings were mostly performed in the morning (approximately 60%), but flowers were also receptive after midday. The crosses

Fig. 11.2. Flowers of some *Sobralia* species used for pollination studies at the Lankester Gardens project. Photos are not to scale. A) *S. danjanzenii* Dressler & Pupulin. B) *S. x gloriana*. C) *S. nutans* Dressler. D) *S. macrantha* Lindley. E) *S. luteola* Rolfe. F) *S. helleri* Hawkes. G) *S. turrialbina* Dressler. H) *S. kerryae* Dressler. I) *S. chrysostoma* Dressler. J) *S. bouchei* Ames & C. Schweinf. K) *S. atropubescens* Ames & C. Schwelnf. L) *S. gloriana* Dressler.

were implemented to observe the capacity of the species to receive and donate pollen to fertilize other plants. Flower buds before anthesis (e.g., about to open) were used. Petals and sepals were carefully detached, then pollinia were extracted with a clamp and placed on the moist (e.g., receptive) stigma of another flower. The receptive stigma was confirmed by the presence of a sticky viscidium. A plastic label noting the registration number of the recipient plant, the donor, the person performing the cross and the date and time was attached to each cross. At all times, data were entered in a digital notebook in Google Docs.

The initial experimental hypothesis in 2020 was that *Sobralia* species would exhibit self-incompatibility, a type of barrier in which the pollen tube cannot extend into the stigma tissues and therefore does not reach the ovules to fertilize them. This process has been well documented in *Brassicaceae* and depends on the interaction of the *S*-allele specific stigma determinant and the pollen determinant (Osabe *et al.*, 2012). However, results suggested that in general, *Sobralia* species can pollinate themselves and other species; that is, capsules are produced that develop and mature normally (Table 11.2).

The results were also graphed to show the median, that is, the typical statistical trend in each species. Likewise, the results of the species were grouped into two categories: the pollination success rate as pollen donors (Fig. 11.3) and the rate as pollen recipients (Fig. 11.4).

The pollination rate of species as pollen donors was 100% for *S. decora*, *S. dorbiguyana*, *S. fenzliana*, *S. kerryae*, *S. macrantha*, and *S. rogersiana*; that is, their pollen was fully capable of self-pollinating and cross-pollinating (Fig. 11.3). The next successful species were *S. atropubescens* and *S. x gloriana* (a showy Panamanian interspecific hybrid), with a rate close to 70%. The pollination rate was close to 50% for *S. chrysostoma*, *S. crispissima*, *S. gloriana*, and *S. luteola* and was less than 50% for *S. leucoxantha*, *S. danjanzenii*, *S. turrialbina*, *S. bouchei*, and *S. helleri*. In other words, in these latter species, despite low pollination rates, healthy and well-developed capsules could be obtained, but many flowers were needed for the crossing. For *S. biflora* 13577 (provisional name) and *S. nutans*, no capsules were obtained.

Table 11.2. Summary of the crosses of *Sobralia* species during the final stage of the project. Author's own table.

Cross pollinations	Number	Species involved
Full reciprocal crosses	6	*S. atropubescens* x *S. gloriana*, *S. chrysostoma* x *S. helleri*, *S. kerryae* x *S. dichotoma*, *S. macrantha* x *S. rosea*, *S. portillae* x *S. chrysostoma*, *S. mutisii* x *S. biflora*
Pollinations across plants of the same species	3	*S. atropubescens*, *S. helleri*
Species that set capsules after self-pollination (88/110: 80% rate)	17	*S. atropubescens*, *S. biflora*, *S. bouchei*, *S. crispissima*, *S. danjanzenii*, *S. decora*, *S. dorbiguyana*, *S. fenzliana*, *S.* x *gloriana*, *S. gloriana*, *S. helleri*, *S. kerryae*, *S. leucoxantha*, *S. luteola*, *S. macrantha*, *S. rogersiana*, *S. rosea*
Species that grew capsules after self-pollination (3/110, 2,7% rate)	3	*S. bouchei*, *S. luteola*, *S. turrialbina*
Species for which flowers dropped after self-pollination (19/110, 17% rate)	7	*S. chrysostoma*, *S. leuconxantha*, *S. crispissima*, *S. nutans*, *S. bouchei*, *S. helleri* cv. *alba*
Stable hybrid grown at the greenhouse after *in vitro* germination	2	*S. kerryae* x *S. dichotoma* and *S.danjanzenii* x *S. dichotoma*
Hybrid that did not survive greenhouse cultivation after *in vitro* germination	1	*S. atropubescens* x *S. gloriana*

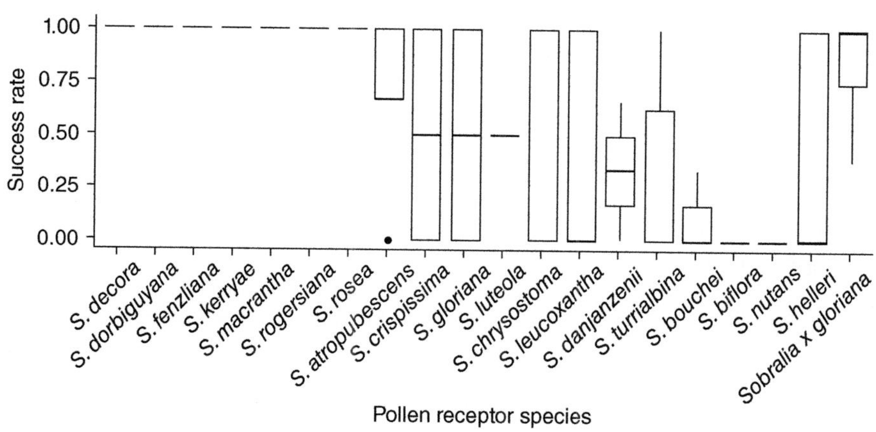

Fig. 11.3. Pollination rate for *Sobralia* species that served as pollen donors (male plants). Each column represents the median, that is, the trend, which in some cases appears at the lower or upper end of the column. Author's own image.

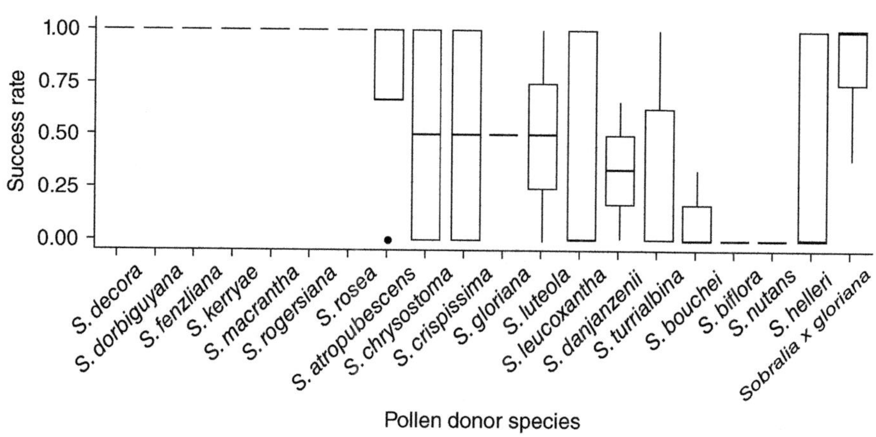

Fig. 11.4. Pollination rate for *Sobralia* species that served as pollen recipients (female plants). Each column represents the median, that is, the trend, which in some cases appears at the lower or upper end of the column. Author's own image.

During the growth of capsules, outright abortion did not occur as in *Phalaenopsis* hybrids, abortion being attributed to defects in the closure of stigmatic cavity (Hsu *et al.*, 2010). Nonetheless, temperature during pollination affects capsule set and seed content in *Phalaenopsis* (Chang *et al.*, 2010), so temperature may affect seed set in some *Sobralia* species as well.

Some species showed 100% ability to receive external pollen and develop capsules (Fig. 11.4). These species were *S. decora*, *S. dorbiguyana*, *S. fenzliana*, *S. kerryae*, *S. macrantha*, and *S. rogersiana*. The rate was 75% for *S. atropubescens* and *Sobralia x gloriana* and was less than 50% for *S. crispissima*, *S. gloriana*, *S. luteola*, *S. danjanzenii*, *S. bouchei*, *S. chrysostoma*, *S. leucoxantha*, *S. turrialbina*, and *S. helleri*. Once again, for *S. biflora* 13577 (provisional name) and *S. nutans*, no capsules could be harvested. Thus, in most *Sobralia* species, flowers may be pollinated with cryogenically stored pollen, as has been done in Thailand with *Phalaenopsis* pollen (Thammasiri, 2007) and at the Costa Rica Technology Institute with *Cattleya* pollen (Alvarado-Ulloa, 2013). This technique favors the conservation and reproduction of endangered species. Unfortunately, we could not determine why *S. nutans* was infertile, although environmental factors may hinder the germination and elongation of pollen tubes. For instance, in tulips, a low temperature (15°C) is required for successful pollen tube elongation and fertilization (Osabe *et al.*, 2012).

11.3 Pollen Viability and Germination Ability in *Sobralia*

Our results suggest that in *Sobralia*, there are no substantial barriers to self-pollination or cross-pollination. Presumptive barriers to pollination were observed in only two species (*S. biflora* and *S. nutans*), although it is not clear why. Preliminary results in two species suggested that pollen viability by tetrazolium staining was moderate to high (36 ± 2% in *S. helleri* and 89 ± 5% in *S. geminata*, n = 600 with three replicates), whereas pollen *in vitro* germination in modified Brewbaker and Kwack medium for 24 h at room temperature (Chen and Fang, 2016) was moderate (31 ± 6% in *S. helleri* and 40 ± 5% in *S. geminata*, n = 1000 with three replicates) (Fig. 11.5A,B). Positive pollen tube extrusion was verified only when tubes were longer than the pollen grain itself (Mora, 2020). Thus, in terms of evolutionary and adaptive capacity, plants from this genus may have ample ability to exchange alleles across populations, with increased chance to generate inter-specific hybrids. Alternatively, the apparent high viability of pollen samples in *Sobralia*

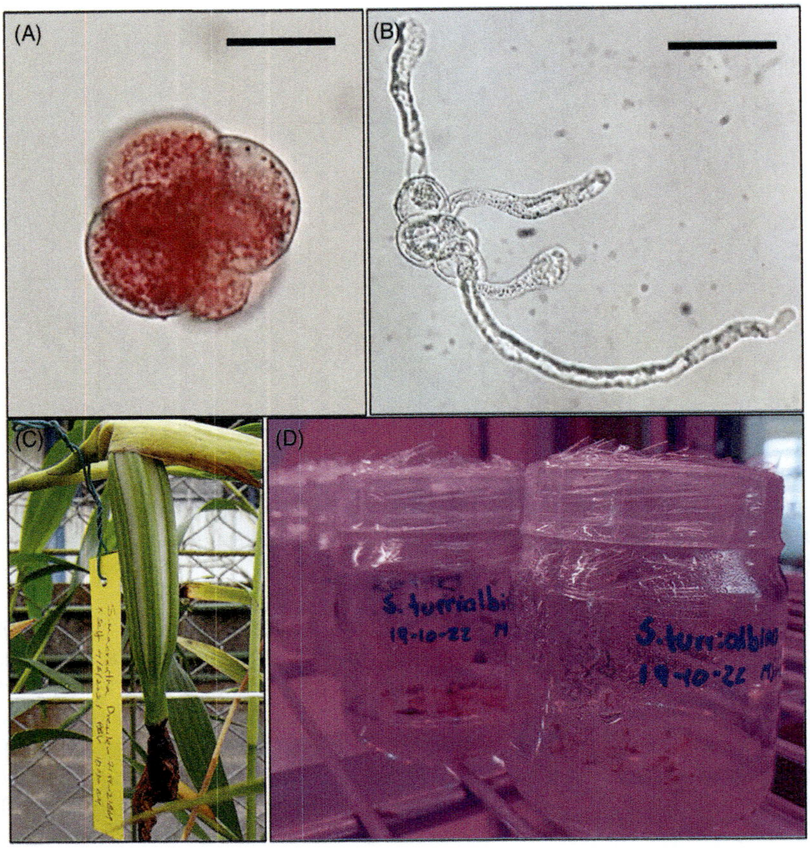

Fig. 11.5. Pollen germination ability and *in vitro* seed germination in *Sobralia.* A) Tetrazolium staining of pollen from *S. geminata*; viable pollen is stained red. As in most orchids, meiotic tetrads in *Sobralia* remained attached after telophase II and continued to develop together as independent microspores during mitosis I and II. B) *In vitro* pollen germination in *S. geminata*. C) Capsule formation in *S. macrantha* after self-pollination. D) *In vitro* seed germination in *S. turrialbina*. Scale bars: A, 10 μm; B, 20 μm. Images A and B are CC-BY-4.0 and reproduced with permission from Mora *et al.* (2018).

suggests that this tissue has great potential for programmed cooling cryopreservation, as suggested by Das *et al.* (2021) for orchid breeding programs. In this technique, orchid pollen is first dehydrated, exposed to cryoprotectants such as glycerin and ethylene glycol, then slowly cooled to –40°C, plunged into liquid nitrogen and finally placed at a suitable temperature, which may range from –20°C to –79°C. Successful post-storage pollen germination has been reported for *Epidendroideae* (Das *et al.*, 2021).

11.4 *In Vitro* Seed Germination in *Sobralia*

For breeding purposes, reproducing *Sobralia* species via *in vitro* seed germination is feasible. For instance, although for *S. klotzscheana*, 80% of seed could be germinated in Murashige and Skoog (MS) medium supplemented with pineapple juice (200 ml l^{-1}) (Salazar-Mercado and Cancino, 2012), we have successfully obtained *in vitro* germination in commercial half-strength MS medium (Duchefa Biochemie, # M0222.0050) for *S. atropubescens*, *S. turrialbina*, *S. helleri*, *S. gloriana*, *S. chrysostoma*, *S. macrantha* (Fig. 11.5C), *S. kerryae*, *S. danjanzenii*, *S. crispissima*, and the hybrids *S. atropubescens* x *S. gloriana*, *S. kerryae* x *S. dichotoma*, and *S. danjanzenii* x *S. dichotoma*. No supplementation with fruit juice was required. In these species and hybrids, healthy-looking seedlings could be grown with no apparent signs of vitrification (Fig. 11.5D). Nonetheless, the main obstacle to propagation from tissue culture is successful acclimation to greenhouse conditions, as in the hybrid *S. atropubescens* x *S. gloriana*, which did not survive. Research into orchid seedling hardening and careful optimization of greenhouse environmental conditions, as done by Venturieri *et al.* (2011), is required to guarantee cultivation into adulthood and to develop a robust pipeline of selected species and hybrids for distribution and display across the country.

In vitro reproduction could facilitate the development of horticulturally promising polyploid lines. Polyploids are individuals whose number of chromosomes is doubled by *in vitro* germination of seed and explants in medium supplemented with molecules that depolymerize microtubules, usually at concentrations of 0.005% (weight/volume) for colchicine, as has been done in *Cymbidium* orchids (Aoyama, 2010), or with the herbicide oryzalin (57 μM) in *Dendrobium* and *Epidendrum* (Miguel *et al.*, 2011). Also, flower buds were smeared with lanolin paste mixed with 0.09–0.13% trifluralin to induce the formation of polyploid male gametes in *Phalaenopsis* (Chen *et al.*, 2021a). Colchicine, oryzalin and trifluralin depolymerize microtubules (Miguel *et al.*, 2011); that is, they disrupt the nucleation of α and β tubulin heterodimers at the plus end of the γ-tubulin ring complex, thus leading to defects in the tethering of the outer kinetochore of chromosomes to the microtubule organizing center during cell division (Hashimoto, 2015; Comai *et al.*, 2017). In polyploids, self-incompatibility is reduced or eliminated because it is believed to alter the expression of genes that regulate the interaction between stigma determinants (*SRK* genes) and pollen determinants (*SP11/SCR* genes) (Osabe *et al.*, 2012). This situation is desirable to propagate materials of high horticultural or scientific value, such as *Sobralia* species, especially those with large flowers such as *S. turrialbina*, *S. gloriana*, *S.* x *gloriana*, *S. atropubescens*, and *S. kerryae*, which is also fragrant.

Eventually, the determinants of accelerated wilting of flowers must be identified at a genetics level. For example, in the model plant *Arabidopsis thaliana*, the *EPHEMERAL1* gene encodes a transcription factor that regulates petal senescence without ethylene (Shibuya, 2018). *EPHEMERAL1* is a key growth regulator in flower senescence, including in orchids of high commercial value such as *Mokara* (intergeneric cross between *Vanda* x *Arachnis*) (Wongjunta *et al.*, 2021). Also, in *Phalaenopsis* orchids, the expression of the genes *FOREVER YOUNG FLOWER 1* and *2* significantly regulates petal wilting, and their silencing in transgenic plants accelerates the wilting process (Chen *et al.*, 2021b). In this way, it might be feasible to (i) characterize in *Sobralia* the expression of these and other genes that have been reported in the literature, (ii) identify step by step how the wilting process operates in *Sobralia*, and (iii) overexpress the coding sequences of these genes to delay the process.

11.5 Conclusions

In summary, results in *Sobralia* species suggest that despite the reported differences in chromosome number (Mora *et al.*, 2018), all species are sexually fertile, pollen germination capacity was verified (Mora, 2020) and seed can be harvested from both self- and cross-pollination. The genomes of *Sobralia* species likely have evolved adaptations for plants to more easily conduct meiotic recombination, elongation of pollen tubes, and efficient fertilization of ovules. However, these three different processes will have to be investigated separately with specialized tools.

Future molecular biology projects might expand our understanding of the immense fertility in the genus *Sobralia* and create possibilities for its horticultural use with *in vitro* cultivation. In the meantime, we will continue to hybridize and grow *Sobralia* species to develop a line of viable commercial varieties. Lankester Gardens has consistently conducted *ex situ* conservation of native species, including *Sobralia* species, at its nurseries, which attests to the need for government-led orchid conservation in tropical countries, as advocated in other countries such as Singapore (Teoh, 2019). These initiatives may have a significant cultural, social and economic impact. We hope that by continuing our research, Costa Rican orchids may be left undisturbed in the forest for locals and tourists to admire for generations.

Acknowledgements

We thank the Vice-Rector for Research and the Vice-Rector for Student Life (University of Costa Rica) for their constant support, and students Jose Mora, Kalani Scott, Carlos Moya, Nelly Murillo, and Evelyn Pérez, who contributed to projects B5A49 and C0244. Special thanks to Grettel Salguero at the Gardens for diagramming and to Ms Laura Smales (Ontario) for careful editing of this manuscript.

References

Alvarado-Ulloa, C. (2013) Cryopreservation of seeds and protocorms of endangered species of the Orchidaceae family: final research project report (original in Spanish). Costa Rica Technology Institute (ITCR).

Aoyama, M. (2010) Chromosome research for orchid breeding. *Acta Horticulturae* 878, 125–131.

Arida, B. L., Scopece, G., Machado, R. M., Moraes, A. P., Forni-Martins, E. and Pinheiro, F. (2021) Reproductive barriers and fertility of two neotropical orchid species and their natural hybrid. *Evolutionary Ecology* 35, 41–64. DOI: 10.1007/s10682-020-10095-5

Bhagwat, N.R., Owens, S.N., Ito, M., Boinapalli, J.V., Poa, P. *et al.* (2021) SUMO is a pervasive regulator of meiosis. *Elife* 10: p.e57720. DOI: 10.7554/eLife.57720

Bolaños-Villegas, P., Chin, S.W. and Chen, F.C. (2008) Meiotic chromosome behavior and capsule setting in *Doritaenopsis* hybrids. *Journal of the American Society of Horticultural Science* 133, 107–116.

Chang, Y.K., Blischak, L.A., Veilleux, R.E. and Iqbal, M.J. (2010) Effect of temperature on gametophytic selection in a *Phalaenopsis* F_1 population. *Euphytica* 171, 251–261. DOI: 10.1007/s10681-009-0040-3

Chen, J.C. and Fang, S.C. (2016) The long pollen tube journey and *in vitro* pollen germination of *Phalaenopsis* orchids. *Plant Reproduction* 29, 179–188. DOI: 10.1007/s00497-016-0280-z

Chen, F.C., Chin, S.W. and Huang, J.Z. (2021a) Novel orchid breeding to meet future market demands. *Acta Horticulturae* 1312, 1–7.

Chen, W.H., Jiang, Z.Y., Hsu, H.F. and Yang, C.H. (2021b) Silencing of *FOREVER YOUNG FLOWER*-Like genes from *Phalaenopsis* orchids promotes flower senescence and abscission. *Plant Cell Physiology* 62, 111–124. DOI: 10.1093/pcp/pcaa145

Comai, L., Maheshwari, S. and Marimuthu, M. P. (2017) Plant centromeres. *Current Opinion in Plant Biology* 36, 158–167.

Das, M. C., Devi, S. D., Kumaria, S. and Reed, B. M. (2021). Looking for a way forward for the cryopreservation of orchid diversity. *Cryobiology* 102, 1–14. DOI: 10.1016/j.cryobiol.2021.05.004

Dressler, R. L. and Pupulin, F. (2014) Two new white-flowered species of *Sobralia* (Orchidaceae) from Costa Rica. *Harvard Papers in Botany* 19, 117–122. https://www.jstor.org/stable/10.2307/harpapbot.19.1.117

Dressler, R. L., Acuña, M. and Pupulin, F. (2016) *Sobralia turrialbina* (Orchidaceae: Sobralieae): long cultivated and now described. *Harvard Papers in Botany*, 21: 251-261. https://www.jstor.org/stable/10.2307/90000326 (accessed 11 July 2025).

Hashimoto, T. (2015) Microtubules in plants. *The Arabidopsis Book/American Society of Plant Biologists* 13: e0179. DOI: 10.1199/tab.0179

Hsu, S.T., Chuang, H.T. and Shen, T.M. (2010) Breeding barriers in red *Phalaenopsis* orchids. *Acta Horticulturae* 878, 145–152.

Marques, A. and Pedrosa-Harand, A. (2016) Holocentromere identity: from the typical mitotic linear structure to the great plasticity of meiotic holocentromeres. *Chromosoma*, 125: 669-681. https://doi.org/10.1007/s00412-016-0612-7

Miguel, T.P. and Leonhardt, K.W. (2011) *In vitro* polyploid induction of orchids using oryzalin. *Scientia Horticulturae* 130, 314–319.

Mora, J. (2020) Analysis of meiotic chromosomes in Costa Rican orchids of the genus Sobralia to establish their viability (original in Spanish). Undergraduate thesis in Agronomy, University of Costa Rica. https://hdl.handle.net/10669/84998 (accessed 11 July 2025).

Mora, J., Albertazzi, S. and Bolaños-Villegas, P. (2018) Meiotic chromosome analysis in tropical orchid genus *Sobralia*. *Cahiers de la Société Française d'Orchidophilie* 9, 117–119. https://hdl.handle.net/10669/76279 (accessed 11 July 2025).

Oliveira, L., Neumann, P., Mata-Sucre, Y., Kuo, Y.T., Marques, A., Schubert, V. and Macas, J. (2024) KNL1 and NDC80 represent new universal markers for the detection of functional centromeres in plants. *Chromosome Research* 32, 3. DOI: 10.1007/s10577-024-09747-x

Osabe, K., Kawanabe, T., Sasaki, T., Ishikawa, R., Okazaki, K., Dennis, E.S., Kazama, T. and Fujimoto, R. (2012) Multiple mechanisms and challenges for the application of allopolyploidy in plants. *International Journal of Molecular Science* 13, 8696–8721.

Salazar-Mercado, S.A. and Cancino, G.O. (2012) Evaluation of the effect of two organic supplements on the *in vitro* germination of native orchids from the province of Pamplona, Colombia (original in Spanish). *Revista Colombiana de Biotecnología* 14, 53–59.

Shibuya, K. (2018) Molecular aspects of flower senescence and strategies to improve flower longevity. *Breeding Science* 68, 99–108. DOI: 10.1270/jsbbs.17081

Tanaka, R. and Kamemoto, H. (1960) Meiotic chromosome behavior in diploid and polyploid Vanda orchid hybrids. *Cytologia* 25, 405–418. DOI: 10.1508/cytologia.25.405

Teoh, E.S. (2019) *Orchids as aphrodisiac, medicine or food*. Springer Nature, Cham, Switzerland. DOI: 10.1007/978-3-030-18255-7

Thammasiri, K. (2007) Cryopreservation of some Thai orchid species. *Acta Horticulturae* 788, 53–62.

Venturieri, G. A. and Arbieto, E. A. M. D. (2011) *Ex-vitro* establishment of *Phalaenopsis amabilis* seedlings in different substrates. *Acta Scientiarum-Agronomy* 33, 495–501. DOI: 10.4025/actasciagron.v33i3.3950

Whitbread, A. L., Dorn, A., Röhrig, S. and Puchta, H. (2021) Different functional roles of RTR complex factors in DNA repair and meiosis in *Arabidopsis* and tomato. *The Plant Journal* 106, 965–977. DOI: 10.1111/tpj.15211

Wongjunta, M., Wongs-Aree, C., Salim, S., Meir, S., Philosoph-Hadas, S. and Buanong, M. (2021) Involvement of ethylene in physiological processes determining the vase life of various hybrids of *Mokara* orchid cut flowers. *Agronomy* 11, 160. DOI: 10.3390/agronomy11010160

12 Intergeneric hybridization of *Holcoglossum* alliance

Kuang-Chao Wu, Hsuan-Ching Lin, Jia-Qi Chang and Mei-Jiuan Li*
Taiwan Seed Improvement and Propagation Station, Ministry of Agriculture, Taichung, Taiwan

Abstract

Orchid genus *Holcoglossum* Schltr. comprises monopodial orchids from Southeast Asia. These are plants with beautiful large flowers and cylindrical leaves. In this report eight *Holcoglossum* species were collected and crossed with allied genera such as *Aerides*, *Phalaenopsis*, *Luisia*, *Luisanda*, *Renades*, *Renanthera*, *Rhynchostylis*, *Vanda*, and *Vandachostylis* to successfully create three intrageneric and 77 intergeneric hybrids. Due to the lack of blue pigmentation in flowers from *Phalaenopsis*, intergeneric *Phalaenopsis* hybrids with big violet flowers have been created as well. Most *Holcoglossum* intergeneric hybrids with *Vanda* and *Rhynchostylis* show desirable traits such as a compact plant size, improved flower size, temperature tolerance, and a short juvenile phase that make them ideal for potted plant commercialization. These novelty orchids created by intergeneric hybridization may adapt well to a new era of energy efficiency, reduced carbon consumption, and extreme climate.

Keywords: *Holcoglossum*, Asian orchids, orchid breeding, blue flowers, thermo tolerance.

12.1 Introduction

Phalaenopsis is the most popular orchid in the global flower industry due to its long-lasting flowers, short juvenile period, and spike forcing all year round. Its breeding development has almost reached a plateau with tremendously diversified varieties in the global market. However, due to climate change, political conflicts, and the advancement of science and technology development around the world, almost every country encounters energy problems. *Phalaenopsis* grows at a suitable temperature of 18–30°C. Vegetative growth often requires heating, and low temperature is also required to induce flowering, resulting in high input of energy consumption and production cost. Thus an increasing demand is called to attention for orchids that can withstand temperature stress and energy-saving cultivation.

The *Holcoglossum* genus was established in 1919 by Rudolf Schlechter. The World Checklist of Monocotyledons now recognizes 23 species (Christenson ,1998; Jin *et al.*, 2004a,b, 2007a,b; Seidenfaden, 1988; Xiang *et al.*, 2012). The species of the genus have small to medium plant size, terete to semi terete leaves, mostly white to pink flowers, and the lip color mixed with red-purple or orange-yellow, orange-red spots, and a long sac-like spur. *Holcoglossum* species display an interesting distribution pattern in

*Corresponding author: ann@tss.gov.tw

 Biotechnology of Orchids and Other Ornamentals (Eds P. Bolaños-Villegas and F-C. Chen)
DOI: 10.1079/9781800627895.0012

Aeridinae by having two diversity centers (Jin, 2005; Fan, *et al.*, 2009; Aver'janov and Averyanova, 2003). One center, like those of most other genera in Aeridinae, is in the tropical border region between China, Vietnam, and Laos. The other distribution is in the temperate alpine region comprising the Hengduan Mountains. This pattern has not been observed in any other genera in the Aeridinae. *Holcoglossum* genus was subdivided into three groups which seem to coincide with the distribution center. (i) The tropical (TC): *H. wangii*, *H. kimballianum*, *H. himalacium*, *H. nagalandensis*, *H. amesianum*, *H. subulifolium* etc. (ii) The alpine (AC): *H. flavescens*, *H. sinicum*, *H. weixiense*, *H. nujiangense*, *H. rupestre*, *H. tsii* etc. (iii) The intermediate (HC): *H. quasipinifolium*, *H. pumilum*, *H. lingulatum*, *H. omeiense* etc. (Zhao *et al.*, 2020).

The monopodial genus *Holcoglossum*, with many heat or cold tolerant species, is genetically closely related to *Vanda*, *Rhynchostylis*, and *Phalaenopsis* in Aeridinae. Novelty orchids created by intergeneric hybridization between these genera are expected to adapt to the new era of energy saving, carbon consumption, and extreme climate.

12.2 Materials and Methods

12.2.1 Intra- and intergeneric hybridization in *Holcoglossum* alliance

In this study, we collected eight species of *Holcoglossum*, including *H. amesianum*, *H. flavescens*, *H. kimballianum*, *H. pumilum*, *H. quasipinifolium*, *H. subulifolium*, *H. tsii*, and *H. wangii* (Table 12.1 P1–P8). Some *Holcoglossum* intergeneric hybrids, such as *Luisia megasepala* x *H. amesianum* (P.1-2), *H. flavescens* x *Vanda* Yi Yuan Biase (P.1-14) were also collected or recorded from other breeders in Taiwan (Table 12.1). We used *Holcoglossum* species and their hybrids to perform intergeneric hybridization with other genera in the Aeridinae, including *Vanda*, *Luisia*, *Rhynchostylis*, *Aerides*, *Paraphalaenopsis*, and *Phalaenopsis*.

12.2.2 Low temperature tolerance test

Ten plants each of *H.* Pink Jenny x *P. amabilis* (Fig. 12.17B), *H.* Pink Jenny x *Rhy. gigantea* (Fig. 12.17C), *H.* Pink Jenny x *V.* Bai Blue (Fig. 12.17D), *P.* Fortune Saltzman (Fig. 12.17A) as control and one or two plants each of (*H. flavescens* x *V. coerulea*) x *P.* Da Shang Sparkle Fire (Fig. 12.18B), (*H. flavescens* x *V. coerulea*) x *P.* Ben Yu Orange (Fig.18C), (*H. flavescens* x *V. coerulea*) x *P.* OX Black Face (Fig. 12.18D), (*H. flavescens* x *V. coerulea*) x *P.* OX Honey (Fig.18E), (*H. flavescens* x *V. coerulea*) x *P.* OX Queen (Fig. 12.18F), (*H. flavescens* x *V. coerulea*) x *P.* Yu Pin Panda (Fig. 12.18G), (*H. flavescens* x *V. coerulea*) x *P.* Yu Pin Easter Island (Fig. 12.18H), (*H. subulifolium x P. philippinensis*) x *P.* Miki Pink Lady (Fig. 12.18I), (*H. subulifolium x V. coerulea*) x *P.* Sogo Yukidian (Fig. 12.18J), (*H. subulifolium x V. coerulea*) x *P.* Sin Bin Honey (Fig. 12.18K), (*H.* Pink Jenny x *V.* Bai Blue) x *P.* Honglin Rose (Fig. 12.18L) were incubated in growth chamber at 5°C and light intensity of 52 μmol m^{-2} s^{-1} for 2 weeks, then moved to room temperature, and chilling injury was recorded for the low temperature tolerance test.

12.3 Results

12.3.1 Intrageneric hybridization in *Holcoglossum*

Three *Holcoglossum* intrageneric hybrids, *H. amesianum* x *H. flavescens* (Fig. 12.1B), *H. quasipinifolium* x *H. wangii* (Fig. 12.9B), *H. wangii* x *H. pumilum* (Fig. 12.14A) were created in this study (Table 12.1). They are the hybrids between tropical clade (*H. amesianum*, *H. wangii*), alpine clade (*H. flavescens*), and intermediate group (*H. pumilum*, *H. quasipinifolium*). Since flowers of *H. pumilum* are very small and plants are not easy to grow under ambient temperature, it is challenging to serve as the mother parent for any hybridization after large numbers of crossing tests. The juvenile phase of *Holcoglossum* hybrids appears to be short. For example, in *H. amesianum* x *H. flavescens*, we pollinated on March 30, 2021, sowed *in vitro* on July 28, 2021 and removed from the flask on May 20, 2022. It started to bloom on March 10, 2023.

12.3.2 Intergeneric hybridization in *Holcoglossum* alliance

We have collected the intergeneric hybrids of *H. amesianum*, *Luisia megasepala* x *H. amesianum*

Table 12.1. *Holcoglossum* species, their intra- and intergeneric hybrids collected and bred in this study. Author's own table.

- *H. amesianum* (P.1[a])
  - *H. amesianum* x *H. flavescens* (P.1-1)
  - [b]*Luisia megasepala* x *H. amesianum* (P.1-2)
    - (*Luisia megasepala* x *H. amesianum*) x *Rhy gigantea* (P.1-2-1)
- *H. flavescens* (P.2)
  - *H. flavescens* x *Renades* Koloa (P.2-1)
  - *H. flavescens* x *Paraphalenopsis labukensis* (P.2-2)
  - *H. flavescens* x *P.* Lioulin Lip (P.2-3)
  - *H. flavescens* x *P. philippinensis* (P.2-4)
  - *H. flavescens* x *P.* Sogo Yukidian (P.2-5)
  - *H. flavescens* x (*P.* Tyinh Shin Unicorn x *P.* Taiwan Leopard) (P.2-6)
    - (*H. flavescens* x *P.* Chian Xen Violin) X *P.* Lioulin Pretty Lip (P.2-7-1)
  - *H. flavescens* x *Renanthera monachica* (P.2-8)
    - (*H. flavescens* x *Rhy. gigantea*) x *P.* Taisuco Stripe Beauty (P.2-9-1)
    - *Holcostylis* Yawi Girl [(*H. flavescens* x *Rhy. gigantea*) x *Rhy. gigantea*] (P.2-9-2)
      - *Holcostylis* Yawi Girl x *P.* Tai Lin Red Angel (P.2-9-2-1)
      - *Holcostylis* Yawi Girl x *P.* Lioulin Violin (P.2-9-2-2)
      - *Holcostylis* Yawi Girl x *V. coerulea* (P.2-9-2-3)
  - *H. flavescens* x *V. coerulea* (P.2-10)
    - (*H. flavescens* x *V. coerulea*) x *P.* Da-Shang Sparkle Fire (P.2-10-1)
    - (*H. flavescens* x *V. coerulea*) x *P.* Ben Yu Orange (P.2-10-2)
    - (*H. flavescens* x *V. coerulea*) x *P.* OX Black Face (P.2-10-3)
    - (*H. flavescens* x *V. coerulea*) x *P.* OX Honey (P.2-10-4)
    - (*H. flavescens* x *V. coerulea*) x *P.* OX Firebird (P.2-10-5)
    - (*H. flavescens* x *V. coerulea*) x *P.* OX Queen (P.2-10-6)
    - (*H. flavescens* x *V. coerulea*) x *P.* Yu Pin Panda (P.2-10-7)
    - (*H. flavescens* x *V. coerulea*) x *P.* Yu Pin Easter Island (P.2-10-8)
  - *H. flavescens* x *V.* Gordon Dillon (P.2-11)
    - (*H. flavescens* x *V.* Gordon Dillon) x *Aerides houlletiana* (P.2-11-1)
    - (*H. flavescens* x *V.* Gordon Dillon) x *P.* Sogo Yukidian (P.2-11-2)
    - (*H. flavescens* x *V.* Gordon Dillon) x (*V.* Bit's Heartthrob x *V.* Dr. Anek) (P.2-11-3)
  - *H. flavescens* x *V. miniata* (P.2-12)
    - (*H. flavescens* x *V. miniata*) x *P.* Heliodor (P.2-12-1)
  - *H. flavescens* x *V.* Tristar Blue (P.2-13)
  - [b]*H. flavescens* x *V.* Yi Yuan Biase (P.2-14)
  - *H. flavescens* x *Vandachostylis* Prapawan (P.2-15)
- *H. kimbalianum* (P.3)
- *H. pumilum* (P.4)
  - *H.pumilum* x *P.* Tyinh Shin Unicorn (P.4-1)
    - [b](*Rhy. gigantea* x *H. pumilum*) x *Rhy. gigantea* (P.4-2-1)
  - *Vandachostylis* Lou Sneary x *H. pumilum* (P.4-3)
- *H. quasipinifolium* (P.5)
  - *H.* Magic Jenny (*H. quasipinifolium* x *H. wangii*) (P.5-1)
    - *Holcosianda* Sweet Jenny (*Luisanda* Red Seagull x *H.* Magic Jenny) (P.5-1-1)
    - *H.* Magic Jenny x *Rhy. gigantea* var. *devariunii* (P.5-1-2)
    - *H.* Magic Jenny x *V.* Princess Mikasa (P.5-1-3)
    - *Holcovanstylis* Blue Jenny (*Vandachostylis* Lou Sneary x *H.* Magic Jenny) (P.5-1-4)
      - *Holcovanstylis* Blue Jenny x *P.* Lioulin Pretty Belle (P.5-1-4-1)
      - *Holcovanstylis* Blue Jenny x *P.* Fuller's Berry (P.5-1-4-2)
      - *Holcovanstylis* Blue Jenny x *P.* OX Honey (P.5-1-4-3)
      - *Holcovanstylis* Blue Jenny x *V.* BangSai Queen (P.5-1-4-4)
      - *Holcovanstylis* Blue Jenny x *V. coerulea* (P.5-1-4-5)
      - *Holcovanstylis* Blue Jenny x *V.* Princess Mikasa (P.5-1-4-6)
      - *Holcovanstylis* Blue Jenny x (*Vandachostylis* Pine Rivers x *V.* Memoria Tan Beng Chye) (P.5-1-4-7)
  - *H. quasipinifolium* x *V. sanderianum* alba (P.5-2)

Continued

Table 12.1. Continued.

- *H. subulifolium* (P.6)
  - *H. subulifolium* x *P. philippinensis* (P.6-1)
    - (*H. subulifolium* x *P. philippinensis*) x *P.* Miki Pink Lady (P.6-1-1)
  - *H. subulifolium* x *P. schilleriana* (P.6-2)
  - *H. subulifolium* x *P.* Tai Lin Red Angel (P.6-3)
  - *H. subulifolium* x *V. coerulea* (P.6-4)
    - (*H. subulifolium* x *V. coerulea*) x *P.* Sogo Yukidian (P.6-4-1)
    - (*H. subulifolium* x *V. coerulea*) x *Pl.* Sin Bin Honey (P.6-4-2)
    - (*H. subulifolium* x *V. coerulea*) x *Rhy. gigantea* (P.6-4-3)
- *H. tsii* (P.7)
- *H. wangii* (P.8)
  - *H. wangii* x *P.* Chian Xen Magpie (P.8-1)
  - *H. wangii* x *P.* Acker's Sweetie (P.8-2)
  - *H. wangii* x *P.* Mount Lip (P.8-3)
  - *H. wangii* x *P.* Sunrise Beautiful Girl (P.8-4)
  - *Holcostylis* Wu Gift (*H. wangii* x *Rhy. gigantea*) (P.8-5)
    - [b]*Holcostylis* Wu Gift x *Rhy. gigantea* (P.8-5-1)
  - *H. wangii* x *V.* Bai Blue (P.8-6)
  - *H. wangii* x *V. miniata* (P.8-7)
  - *H.* Pink Jenny (*H. wangii* x *H. pumilum*) (P.8-8)
    - *H.* Pink Jenny x *P. amabilis* (P.8-8-1)
    - *H.* Pink Jenny x *P.* Chian Xen Magpie (P.8-8-2)
    - *H.* Pink Jenny x *P.* Lioulin Barbie (P.8-8-3)
    - *H.* Pink Jenny x *P.* Lioulin Pure Lip (P.8-8-4)
    - *H.* Pink Jenny x *P.* Luchia Lip (P.8-8-5)
    - *H.* Pink Jenny x *P.* Luchia Pink (P.8-8-6)
    - *H.* Pink Jenny x *Rhy. gigantea* (P.8-8-7)
    - *H.* Pink Jenny x *V.* Bai Blue (P.8-8-8)
      - (*H.* Pink Jenny x *V.* Bai Blue) x *P.* Heliodor (P.8-8-8-1)
      - (*H.* Pink Jenny x *V.* Bai Blue) x *P.* Honglin Rose (P.8-8-8-2)
      - (*H.* Pink Jenny x *V.* Bai Blue) x *P.* Tai Lin Red Angel (P.8-8-8-3)
    - *H.* Pink Jenny x *V.* Dr. Anek (P.8-8-9)
    - *H.* Pink Jenny x *V.* Manuvadee (P.8-8-10)
    - *H.* Pink Jenny x *V.* Virginia Vercillo (P.8-8-11)
    - *H.* Pink Jenny x *Vandachostylis* Lou Sneary (P.8-8-12)

[a]P.1:Pedigree no. 1
[b]*Holcoglossum* intergeneric hybrids recorded from other nurseries in Taiwan.

(Fig. 12.1C,D) and crossed with *Rhy. gigantea* successfully (Fig. 12.1E). The resulting hybrid is characterized by the short spike, long internodes from *Luisia* and more round-shape flowers from *Rhynchostylis*.

We have obtained many immediate offsprings of *H. flavescens* from crossing with other genera, such as *Renades*, *Renanthera*, *Paraphalaenopsis*, *Vandachostylis* (Fig. 12.2), *Phalaenopsis* (Fig. 12.3), *Rhynchostylis* (Fig. 12.4), *Vanda* (Fig. 12.5). Due to the white flowers of *H. flavescens*, most offsprings develop light colored flowers.

Nonetheless, the flower color of *H. flavescens* x *V. coerulea* ranged from white to pink with pink or violet lips (Fig. 12.5A–C) and flowers of *H. flavescens* x *V. miniata* (Fig. 12.6E) and *H. flavescens* x *V.* Yi Yuan Biase (Fig. 12.6H) are bright yellow and orange from the strong colored pollen parents. The offsprings of *H. flavescens* are fertile and can be crossed with other genera in the *Aerides* (Fig. 12.6B), *Phalaenopsis* (Figs 12.3I, 12.4A,E,H, 12.5D–J, and 12.6C,F), *Rhynchostylis* (Fig. 12.4C), and *Vanda* (Figs 12.4G and 12.6H). The next generation of intergeneric hybrids show a dark flower color, such as in (*H. flavescens* x *P.* Chian Xen Violin) x *P.* Lioulin Pretty Lip (Fig. 12.3I), *Holcostylis* Yawi Girl x *P.* Tai Lin Red Angel (Fig. 12.4E), (*H. flavescens* x *V. coerulea*) x *P.* OX Black Face (Fig. 12.5D–G), and in (*H. flavescens* x *V.* Gordon Dillon) x (*V.* Bit's

Fig. 12.1. *Holcoglossum amesianum* and several intra and intergeneric hybrids. A) *H. amesianum*, P.1. B) *H. amesianum* x *H. flavescens*, P.1-1. C, D) *Luisia megasepala* x *H. amesianum,* P.1-2. E) (*L. megasepala* x *H. amesianum*) x *Rhy. gigantea,* P.1-2-1. Author's own image.

Heartthrob x *V.* Dr. Anek) (Fig. 12.6D), but also may grow big flowers as in (*H. flavescens* x *V. coerulea*) x *P.* OX Black Face (Fig. 12.5D–G) which are also violet, a trait that is exceedingly rare in the *Phalaenopsis* market.

As a note of caution, *H. kimbalianum* (Fig. 12.7A) was not used to create any progeny because of the commercial disadvantage of growing elongated stems as potted orchids.

In *H. pumilum*, despite expressing desirable traits such as multiple spikes and multiple flowers that are in demand for the breeding of potted orchids, its tiny flowers and difficult cultivation makes it hard to adopt it as a parent for breeding elite hybrids. Luckily, we have bred one direct progeny using *H. pumilum* as a seed parent crossed with *P.* Tying Shin Unicorn (Fig. 12.8C,D). The flowers of another cross, *Vandachostylis* Lou Sneary x *H. pumilum* (Fig. 8F–H), are either white, pink or violet and release a pleasant fragrance. Another breeder in Taiwan has managed to breed an interesting hybrid involving (*Rhy. gigantea* x *H. pumilum*) x *Rhy. gigantea* (Fig. 12.8E) and its grex name is *Holcostylis* TLDC Fuchsia Stairway. This plant is compact with a substantial number of flowers.

We have also used the intrageneric hybrid *H. quasipinifolium* x *H. wangii* (Fig. 12.9B) as a bridge to cross with genus *Luisanda* (Fig. 12.9C), *Rhynchostylis* (Fig. 9E,F), and *Vanda* (Fig. 12.9GH) successfully. For fragrance and violet flower color breeding, we further used *Holcovanstylis* Blue Jenny (Fig. 12.10A) as seed parent to cross with *Phalaenopsis* (Fig. 12.10B,C) and *Vanda* (Fig. 12.10D–J). However, *Phalaenopsis* hybrids (Fig. 12.10B,C) show the segregation of the long leaf trait inherited from the mother parent. Some *Holcovanstylis* Blue Jenny x *V.* Bangsai

Fig. 12.2. *H. flavescens* and its intergeneric hybrids. A) *H. flavescens*, P.2. B, C) *H. flavescens x Renades* Koloa, P.2-1. D) *H. flavescens x Renanthera monachica*, P.2-8. E) *H. flavescens x Paraphalenopsis labukensis*, P.2-2. F, G) *H. flavescens x Vandachostylis* Prapawan, P.2-15. Author's own image.

Fig. 12.3. *Holconopsis* hybrids bred from *H. flavescens*. A) *H. flavescens* x *P.* Lioulin L Lip, P.2-3, B–D) *H. flavescens* x *P. philippinensis*, P.2-4. E) *H. flavescens* x *P.* Sogo Yukidian, P.2-5. F–H) *H. flavescens* x (*P.* Tying Shin Unicorn x *P.* Taiwan Leopard), P.2-6. I) (*H. flavescens* x *P.* Chian Xen Violin) x *P.* Lioulin Pretty Lip, P.2-7-1. Author's own image.

Queen offsprings (Fig. 12.10D,E) may be marketed as ideal potted orchids with a strong fragrance and a short juvenile period of 1.5 years to flowering.

We crossed *H. subulifolium* with genera *Phalaenopsis* and *Vanda*. The immediate intergeneric offsprings are characterized by a long upright inflorescence, long leaves, and broadly ovate,

Fig. 12.4. *Holcostylis* hybrid (*H. flavescens* x *Rhy. gigantea*) from *H. flavescens* and the next generation hybrids with *Phalaenopsis* or *Vanda*. A, B) (*H. flavescens* x *Rhy. gigantea*) x *P.* Taisuco Stripe Beauty, P.2-9-1. C, D) (*H. flavescens* x *Rhy. gigantea*) x *Rhy. gigantea* (*Holcostylis* Yawi Girl), P.2-9-2. E) *Holcostylis* Yawi Girl x *P.* Tai Lin Red Angel, P.2-9-2-1. F) *Holcostylis* Yawi Girl x *P.* Lioulin Topaz, P.2-9-2-2. G) *Holcostylis* Yawi Girl x *V. coerulea*, P.2-9-2-3. Author's own image.

entire margined lip petal (Fig. 12.11C,D,F,G). Comparing *H. subulifolium* x *V. coerulea* with *H. flavescens* x *V. coerulea* (Fig. 12.11G), they showed the key traits of their seed parents respectively, bigger plant size and long inflorescence from *H. subulifolium*, short plant size and inflorescence but round flower shape from *H. flavescens*. We also produced next generation from *H. subulifolium* x *V. coerulea* crossing with *Phalaenopsis* (Fig. 12.18J,K) and *Rhy. gigantea* (Fig. 12.11I).

Regarding *Holcoglossum tsii*, this is a particularly good species with big pink flowers and a noticeable fragrance (Fig. 12.7B). We have recently collected this species, but it has not been reproduced or crossed by us yet.

A useful species is *H. wangii*, which belongs to the tropical clade of *Holcoglossum* and are cultivated well in Taiwan. Many immediate offspring crossed with *Phalaenopsis* (Fig. 12.12), *Rhynchostylis* and *Vanda* (Fig. 12.13) were produced much easier than other species in *Holcoglossum*. The long leaf character of *Holconopsis* (Fig. 12.12C,F,G,H) and *Vandoglossum* (Fig. 12.13E,F) is from seed parent, *H. wagii*. Leaves of *H. wangii* x *V.* Bai Blue (P.8-5) are much longer than *H.* Pink Jenny (*H. wangii* x *H. pumilum*) x *V.* Bai Blue (Fig. 12.13E) and showed that *H. pumilum* shortened leaf length and contributed to the violet flower color. Some variation is observed in *H. wangii* x *Rhy. gigantea (Holcostylis* Wu Gift) (Fig. 12.13A–C) hybrids, with flowers of different size, and marked differences in inflorescence length, leaf length and leaf width. Other breeders in Taiwan have bred *Holcostylis* Wu Gift x *Rhy. gigantea* (Fig. 12.13D).

We have used the intrageneric hybrid *H. wangii* x *H. pumilum* (Fig. 12.14A) as a bridge to successfully cross with genera *Phalaenopsis* (Fig. 12.14), *Rhynchostylis*, *Vandachostylis* (Fig. 12.15) and *Vanda* (Fig. 12.16). Some of the progenies as in the cases of *H.* Pink Jenny x *P. amabilis* (Fig. 14B,C), *H.* Pink Jenny x *P.* Luchia Lip (Fig. 12.14I), and *H.* Pink Jenny x *Rhy. gigantea* (Fig. 12.15A,D,E), even showed a multi-spike trait derived from *H. pumilum*. Compared with a large plant of *Rhy. gigantea*, a 5-year-old plant of *H.* Pink Jenny x *Rhy. gigantea* is still truly short with only five spikes (Fig. 12.15E).

The plant sizes (about 20 cm in height) of most *H.* Pink Jenny x *V.* Manuvadee seedlings (Fig. 12.16M,N) are shorter than other *Vandoglossum*

Fig. 12.5. *Vandoglossum* hybrid (*H. flavescens* x *V. coerulea*) from *H. flavescens* and the next generation hybrids with *Phalaenopsis*. A–C) *H. flavescens* x *V. coerulea*, P.2-10. D–G) (*H. flavescens* x *V. coerulea*) x *P.* OX Black Face, P.2-10-3. H) Left to right: (*H. flavescens* x *V. coerulea*) x *P.* OX Black Face, (*H. flavescens* x *V. coerulea*) x *P.* Yu Pin Panda, two plants of (*H. flavescens* x *V. coerulea*) x *P.* Ben Yu Orange. I) (*H. flavescens* x *V. coerulea*) x *P.* OX Honey, P.2-10-4. J) (*H. flavescens* x *V. coerulea*) x *P.* OX Firebird, P.2-10-5. Author's own image.

(about 40cm in height), *H.* Pink Jenny x *V.* Virginia Vercillo (Fig. 12.16O,P), *H.* Pink Jenny x *V.* Bai Blue (Fig. 12.16A–D) and *H.* Pink Jenny x *V.* Dr. Anek (Fig. 12.16I–K), due to the xenia effect (e.g. influence of pollen on plant development). The hybrid of *H.* Pink Jenny x *V.* Bai Blue with *Holcoglossum* pedigree is much shorter than that of *Vandachostylis* Lou Sneary x *V.* Bai Blue (Fig. 12.16N). The progeny of *H.* Pink Jenny x *V.* Bai Blue combined advantageous traits such as a short juvenile phase, and cold tolerance from *H. pumilum*, big flower from *H. wangii*, and blue flower from *Vanda*. Subsequently we performed further hybridizations with *Phalaenopsis* to create (*H.* Pink Jenny x *V.* Bai Blue) x *P.* Heliodor (Fig. 12.16E) and (*H.* Pink Jenny x *V.* Bai Blue) x *P.* Tailin Red Angel (Fig. 12.16F,G), characterized with ideal violet big flowers and shorter leaves.

12.4 Low Temperature Tolerance

When incubated in a growth chamber at 5°C for 2 weeks, all leaves of *P.* Fortune Saltzman got chilling injury totally (Fig. 12.17A); and a few young leaves of *H.* Pink Jenny x *P. amabilis* (Fig. 12.17B), *H.* Pink Jenny x *V.* Bai Blue (Fig. 12.17D) got chilling injury too. (*H.* Pink Jenny x *Rhy. gigantea* (Fig. 12.17C) few necrotic spots appeared on

Fig. 12.6. *Vandoglossum* hybrids from *H. flavescens* and the next generation bred with *Aerides, Phalaenopsis* and *Vanda*. A) *H. flavescens* x *V.* Gordon Dillon, P.2-11. B) (*H. flavescens* x *V.* Gordon Dillon) x *Aerides houlletiana*, P.2-11-1. C) (*H. flavescens* x *V.* Gordon Dillon) x *P.* Sogo Yukidian, P.2-11-2. D) (*H. flavescens* x *V.* Gordon Dillon) x (*V.* Bit's Heartthrob x *V.* Dr. Anek), P.2-11-3. E) *H. flavescens* x *V. miniata*, P.2-12. F) (*H. flavescens* x *V. miniata*) x *P.* Heliodor, P.2-12-1. G) *H. flavescens* x *V.* Tristar Blue, P.2-13. H) *H. flavescens* x *V.* Yi Yuan Biase, P.2-14. Author's own image.

Fig. 12.7. A) *H. kimbalianum,* P.3. B) *H. tsii,* P.7 collected in this study. Author's own image.

young and old leaves randomly. *Phalaenopsis* and *Vanda* are not tolerant to low temperatures, except for (*H.* Pink Jenny x *V.* Bai Blue) x *P.* Honglin Rose (Fig. 12.18A, L), thus all intergeneric hybrid plants of *Phalaenopsis* and *Holcoglossum* (Fig. 12.18) are much more tolerant to low temperatures.

12.5 Discussion

Holcoglossum are closely related to many genera in the Aeridinae (Zou et al., 2015). As early as 1944, intergeneric hybrids between *H. kimballianum* and *Papilionanthe* Miss Joaquim were crossed

Fig. 12.8. *H. pumilum* and its inter-generic hybrids. A, B) *H. pumilum,* P.4. C, D) *H.pumilum* x *P.* Tying Shin Unicorn, P.4-1. E) (*Rhy. gigantea* x *H.pumilum*) x *Rhy. gigantea,* P.4-2-1. F–H) *Vandachostylis* Lou Sneary x *H. pumilum,* P.4-3. Author's own image.

Fig. 12.9. *H. quasipinifolium* and the intra-, inter- generic hybrids. A) *H. quasipinifolium,* P.5. B) *H.* Magic Jenny (*H. quasipinifolium* x *H. wangii*), P.5-1. C) *Holcosianda* Sweet Jenny (*Luisanda* Red Seagull x *H.* Magic Jenny), P.5-1-1. D) *H. quasipinifolium* x *V. sanderianum var. alba,* P.5-2. E, F) *H.* Magic Jenny x *Rhy. gigantea,* P.5-1-2. G, H) *H.* Magic Jenny x *V.* Princess Mikasa, P.5-1-2. Author's own image.

and registered as *Holcanthe* Any Frederique Restrepo. In this study, *Holcoglossum* intergeneric hybrids were bred with *Vanda, Luisia, Aerides, Renanthera, Rhynchostylis,* and *Phalaenopsis* successfully. Offsprings from intergeneric hybridization between two genera, three genera, and four genera are available (Table. 12.1), with highest level at fourth generation. This work reveals

Fig. 12.10. *Holcovanstylis* hybrid from *H. quasipinifolium* and the next generation bred with *Phalaenopsis, Vanda* and *Vandachostylis*. A) *Holcovanstylis* Blue Jenny (*Vandachostylis* Lou Sneary x *H.* Magic Jenny), P.5-1-4. B) *Holcovanstylis* Blue Jenny x *P.* Lioulin Pretty Belle, P.5-1-4-1. C) *Holcovanstylis* Blue Jenny x *P.* Fuller's Berry, P.5-1-4-2. D, E) *Holcovanstylis* Blue Jenny x *V.* BangSai Queen, P.5-1-4-4. F) *Holcovanstylis* Blue Jenny x *V. coerulea*, P.5-1-4-5. G) *Holcovanstylis* Blue Jenny x *V.* Princess Mikasa, P.5-1-4-6. H–J) *Holcovanstylis* Blue Jenny x (*Vandachostylis* Pine Rivers x *V.* Memoria Tan Beng Chye), P.5-1-4-7. Author's own image.

the possibility of intergeneric hybridization across multiple genera, with fertile offsprings. Indeed, intergeneric hybridization is rather challenging, because of the low fruit setting rate, low pod maturity rate, and low germination rate. Offsprings are obtained by crossing a large number of plants in a specific environment in this study. One drawback is some of the progenies develop either malformed leaves or misshapen flowers (Fig. 12.16L). Hot or cold tolerant *Holcoglossum* species were subdivided into three groups: (i) hot tolerant species, the tropical (TC); (ii) cold tolerant species, the alpine (AC); and (iii) the intermediate (HC). The intergeneric hybrids from tropical type *H. wangii* and alpine type *H. flavescens* (Fig. 12.13F) crossed with *Vanda miniata* were exposed to elevated temperature (25–30°C) to assess shelf life during the month of May in Taiwan. The flowers of the tropical type hybrid lasted longer for one and a half months, but the alpine-type hybrid (Fig. 12.13G) withered quickly. It showed that

Fig. 12.11. *H. subulifolium* and its inter- generic hybrids. A, B) *H. subulifolium,* P.6. C, D) *H. subulifolium* x *P. philippinensis*, P.6-1. E) *H. subulifolium* x *P. schilleriana*, P.6-2. F) *H. subulifolium* x *P.* Tai Lin Red Angel, P.6-3. G) Left: *H. subulifolium* x *V. coerulea*, P.6-4 and Right: *H. flavescens* x *V. coerulea*, P.2-10. H) *H. subulifolium* x *V. coerulea*, P.6-4. I) (*H. subulifolium* x *V. coerulea*) x *Rhy. gigantea*, P.6-4-3. Author's own image.

Fig. 12.12. *H. wangii* and its intergeneric hybrids bred with *Phalaenopsis*. A, B) *H. wangii,* P.8. C, D) *H. wangii* x *P.* Chian Xen Magpie, P.8-1. E) *H. wangii* x *P.* Acker's Sweetie P.8-2. F) *H. wangii* x *P.* Mount Lip, P.8-3. G, H) *H. wangii* x *P.* Sunrise Beautiful Girl, P.8-4. Author's own image.

these *Holcoglossum* intergeneric hybrids with the TC group makes the offspring comparatively more tolerant to hot temperature than crosses with the AC group. But even when crossing *Phalaenopsis* directly with the TC group the intergeneric hybrids also showed cold tolerance (Fig. 12.18I–K). A cold-sensitive intergeneric hybrid, (*H.* Pink Jenny x *V.* Bai Blue) x *P.* Honglin Rose (Fig. 12.18L) also was observed. The intergeneric hybrids between H. Pink Jenny and

Fig. 12.13. *Holcostylis* and *Vandoglossum* hybrids from *H. wangii*. A–C) *H. wangii* x *Rhy. gigantea* (*Holcostylis* Wu Gift), P.8-5. D) *Holcostylis* Wu Gift x *Rhy. gigantea*, P.8-5-1. E) Left: *H.* Pink Jenny x *V.* Bai Blue, P.8-8-8; Right: *H. wangii* x *V.* Bai Blue, P.8-6. F, G) Left: *H. wangii* x *V. miniata*, P8-7; Right: *H. flavescens* x *V. miniata*, P. 2-12. Author's own image.

Fig. 12.14. Intrageneric hybrid *from H. wangii* and its progeny bred with *Phalaenopsis*. A) *H.* Pink Jenny (*H. wangii* x *H. pumilum*), P.8-8. B, C) *H.* Pink Jenny x *P. amabilis,* P.8-8-1. D, E) *H.* Pink Jenny x *P.* Chian Xen Magpie, P.8-8-2. F) *H.* Pink Jenny x *P.* Lioulin Barbie, P.8-8-3. G) *H.* Pink Jenny x *P.* Lioulin Pure Lip, P.8-8-4. H) *H.* Pink Jenny x *P.* Luchia Pink, P.8-8-5. I, J) *H.* Pink Jenny x *P.* Luchia Lip, P.8-8-6. Author's own image.

Fig. 12.15. The offsprings of intra-generic hybrid *H.* Pink Jenny bred with *Rhynchostylis* and *Vandachostylis*. A–E) *H.* Pink Jenny x *Rhy. gigantea,* P.8-8-7. F–I) *H.* Pink Jenny x *Vandachostylis* Lou Sneary, P.8-8-12. Author's own image.

Fig. 12.16. The next generation of intrageneric hybrid *H.* Pink Jenny crossed with *Vanda*. A–D) *H.* Pink Jenny x *V.* Bai Blue, P.8-8-8 and two third generation bred with *Paenopsis.* E) (*H.* Pink Jenny x *V.* Bai Blue) x *P.* Heliodor, P.8-8-8-1. F. G) (*H.* Pink Jenny x *V.* Bai Blue) x *P.* Tai Lin Red Angel, P.8-8-8-3. H) Left: *H.* Pink Jenny x *V.* Bai Blue, P.8-8-8; Right: *H.* Pink Jenny x *V.* Dr. Anek, P.8-8-9. I–L) *H.* Pink Jenny x *V.* Dr. Anek, P.8-8-9. M) *H.* Pink Jenny x *V.* Manuvadee, P.8-8-10. N) Left: *H.* Pink Jenny x *V.* Manuvadee; Middle: *H.* Pink Jenny x *V.* Bai Blue; Right: *Vandachostylis* Lou Sneary x *V.* Bai Blue. O, P) *H.* Pink Jenny x *V.* Virginia Vercillo, P.8-8-11. Author's own image.

Fig. 12.17. Orchid plants incubated in growth chamber at 5°C for 2 weeks to check chilling tolerance. A) *P.* Fortune Saltzman. B) *H.* Pink Jenny x *P. amabilis*, P.8-8-1. C) *H.* Pink Jenny x *Rhy. gigantea*, P.8-8-7. D) *H.* Pink Jenny x *V.* Bai Blue, P.8-8-8. Author's own image.

Fig. 12.18. *P.* Fortune Saltzman vs individual intergeneric hybrid plants of *Holcoglossum* with *Phalaenopsis* incubated in growth chamber at 5°C for 2 weeks to check chilling tolerance. A) Intergeneric hybrids of *Holcoglossum* with *Phalaenopsis*. B) (*H. flavescens* x *V. coerulea)* x *P.* Da-Shang Sparkle Fire, P.2-10-1. C) (*H. flavescens* x *V. coerulea*) x *P.* Ben Yu Orange, P.2-10-2. D) (*H. flavescens* x *V. coerulea)* x *P.* OX Black Face, P.2-10-3. E) (*H. flavescens* x *V. coerulea)* x *P.* OX Honey, P.2-10-4. F) (*H. flavescens* x *V. coerulea)* x *P.* OX Queen, P.2-10-6. G) (*H. flavescens* x *V. coerulea)* x *P.* Yu Pin Panda, P.2-10-7. H) (*H. flavescens* x *V. coerulea)* x *P.* Yu Pin Easter Island, P.2-10-8. I) (*H. subulifolium* x *P. philippinensis*) x *P.* Miki Pink Lady, P.6-1-1. J) (*H. subulifolium* x *V. coerulea*) x *P.* Sogo Yukidian, P.6-4-1. K) (*H. subulifolium* x *V. coerulea*) x *P.* Sin Bin Honey, P.6-4-2. L) (*H.* Pink Jenny x *V.* Bai Blue) x *P.* Honglin Rose, P.8-8-8-2. Author's own image.

Phalaenopsis, *Rhynchotylis*, and *Vanda* are cold tolerant (Fig. 12.17). This suggests they can be cultivated without heating in temperate zones globally.

The plant size of *Holcoglossum* species is small to medium. *H. pumilum* is a tiny endemic species of Taiwan found at an altitude of about 1200–2300 meters. The intergeneric hybrids between *H. pumilum*, *Phalaenopsis*, and *Rhynchostylis* are shorter plants and have bigger flowers when compared to *H. pumilum*. Thus, *H. pumilum* is the ideal breeding germplasm to confer a small compact plant size but not reducing flower size drastically (Fig. 12.8). Most *Holcoglossum* intergeneric hybrids such as *Vandoglossum* and *Holcostylis* feature a compact plant size and big flowers, which makes them suitable as commercial pot orchids. Short juvenile phase is another advantage of intergenic hybrids of *Holcoglossum*. The *H. amesianum* x *H. flavescens* hybrid required only 2 years from pollination of the parents to flowering of the progenies. Compared with the long juvenile period of *Vanda*, we observed the intergeneric hybrid (*H. flavescens* x *V.* Gordon Dillon) x (*V.* Bit`s Heartthrob x *V.* Dr. Anek) (Fig. 12.6D) bloomed in just 18 months after leaving the flask. These novelty orchids that were created by intergeneric hybridization, may adapt better to heat stress in a time of energy savings at greenhouses and extreme climate, enabling efficiency and environmental protection.

References

Aver'janov, L.V. and Averyanova, A.L. (2003) Updated checklist of the orchids of Vietnam. Vietnam National University Publishing House, Hanoi.

Christenson, E. (1998) Two new species of *Holcoglossum* Schltr. (Orchidaceae: Aeridinae) from China. *Lindleyana* 13, 121–1244.

Fan, J., Qin, H.N., Li, D.Z. and Jin, X.H. (2009) Molecular phylogeny and biogeography of *Holcoglossum* (Orchidaceae: Aeridinae) based on nuclear ITS, and chloroplast trnL-F and matK. *Taxon* 58, 849–861.

Jin, X.H. (2005) Generic delimitation and a new infrageneric system in the genus *Holcoglossum* (Orchidaceae: Aeridinae). *Botanical Journal of the Linnean Society* 149, 465–468.

Jin, X.H., Chen, S.C. and Li, D.Z. (2007a) *Holcoglossum nujiangense* (Orchidaceae: Aeridinae)–a new species and its pollination system. *Nordic Journal of Botany* 25, 125–128.

Jin, X.H., Zhang, T., Gu, Z.J. and Li, D.Z. (2007b) Cytological studies on the genus Holcoglossum (Orchidaceae). *Botanical Journal of the Linnean Society* 154, 283–288.

Jin, X.H., Chen, S.C., Qin, H.N., Zhu, G.H. and Laiping, G.S. (2004a) A new species of Holcoglossum (Orchidaceae) from China. *Novon* 14, 178–179.

Jin, X.H., Qin, H.N. and Chen, S.C. (2004b) A new species of *Holcoglossum* (Orchidaceae: Aeridinae) from China. *Kew Bulletin* 59, 633–635.

Seidenfaden, G. (1988) Orchid genera in Thailand: 14. Fifty-nine vandoid genera. *Opera Botanica* 95, 1–398.

Xiang, X.G., Li, D.Z., Jin, X.H., Hu, H., Zhou, H.L. and Jin, W.T. (2012) Monophyly or paraphyly–the taxonomy of Holcoglossum (Aeridinae: Orchidaceae). *PLoS One* 7: e52050.

Zhao, J., Zhou, P., Li, X., Zhang, L., Jin, X. and Xiang, X. (2020) Temporal and Spatial Pattern of Holcoglossum Schltr. (Orchidaceae), an East Asian Endemic Genus, Based on Nuclear and Chloroplast Genes. *Frontiers in Ecology and Evolution* 8, 245.

Zou, L.H., Huang, J.X., Zhang, G.Q., Liu, Z.J. and Zhuang, X.Y. (2015) A molecular phylogeny of Aeridinae (Orchidaceae: Epidendroideae) inferred from multiple nuclear and chloroplast regions. *Molecular Phylogenetics and Evolution* 85, 247–254. DOI: 10.1016/j.ympev.2015.02.014.

13 Modern Approaches to the Hybridization of Roses

Ping Lim*
Ornamental Plant Research, Altman Plants San Diego, California, USA

Abstract
Due to growing concerns about global climate change and synthetic chemicals affecting the environment, maintaining natural beauty may seem increasingly difficult. Modern rose breeders are making progress in developing roses capable of enduring these stresses while enhancing the planet with minimal harm. The current trend in landscaping and gardening involves using fewer chemicals for a more ecologically friendly approach. This chapter examines commercial-led rose hybridizer advancements in creating new varieties of roses that can endure environmental challenges.

Keywords: Cross hybridization, *Rosa* species, floriculture, genetics, ploidy, back cross, disease, hardiness, own root.

13.1 Introduction

Rose breeding began in Europe in the mid-19th century and expanded significantly in the mid-20th century following the implementation of US plant patent law in 1930. Despite over a century of research into rose genetics, knowledge remains less comprehensive compared to key agricultural food crops. The parentages of the first modern hybrid tea rose “La France” from 1867 are still unknown. This lack of documented parentage data has led to an emphasis on empirical breeding practices.

The rose is not only the world’s most popular flower, but it is also the most economically valuable floral crop. According to the 2014 United States Department of Agriculture Census, the total value of cut roses sold in the USA amounted to \$22,165,000, while an estimated 36,632,061 garden roses were valued at \$203,558,000. This growing economic significance continues to impact the cut flower and landscape gardening industries. Currently, more than 40,000 cultivars of roses are listed in *Modern Rose 12* (Young, 2007), with numerous new varieties introduced globally each year. This abundance of genetic resources offers substantial opportunities for enhancement, including developing hardier roots, improving health, and integrating fragrance, beauty, and essential horticultural principles.

13.2 Materials and Methods

The *Modern Rose 12* catalog (Young, 2007) provides a list of over 40,000 cultivars of modern, old, and species roses, which presents significant

*Corresponding author: pinglim8@msn.com

 Biotechnology of Orchids and Other Ornamentals (Eds P. Bolaños-Villegas and F-C. Chen)
DOI: 10.1079/9781800627895.0013

genetic diversity for breeding purposes. Conventional hand pollination is used to gather desired gene pools from vigorous *Rosa* species and their hybrids, incorporating novel traits such as color, form, and fragrance. Additionally, pollen management, hip set management, cultural practices, and record keeping are routine activities. Budget considerations include seasonal part-time assistance, and greenhouse and field space equipped with irrigation systems, for a minimum 6-year term program.

13.3 Objectives

Creating a rose that requires no chemical dependence involves ensuring several attributes: compact size, hardiness, health, beauty, fragrance, repeat blooming with prolific blooming power, and ease of own root propagation. The breeding effort may involve accumulating the desired bloodline for next-generation hybridization.

13.4 Planning

The strategy for matching parents relies upon the knowledge of each parental pedigree and experience. It is critical to engage seed-setting female parents to match with other desired genes from male parents. Increasing intensity via backcrossing and selfing might be needed over a few generations. The more crosses made the chances of reaching the breeding goals.

The process of breeding and confirming an elite seedling suitable for commercialization typically spans a minimum of 6 years, as outlined in the following phases (Table 13.1):

- Phase 1: Set up and prepare parent plants and conduct pollinations.
- Phase 2: Germinate seeds and select seedlings.
- Phase 3: Propagate selected seedlings.
- Phase 4: Perform secondary selection and distribute to various geographical trial locations.
- Phase 5: Conduct final selection and prepare for release with production schedules.
- Phase 6: Register, apply for patents, and commercialize.

Hiring marketing agencies can offer specialized knowledge to speed up the launch of new products into broader markets, including forming international partnerships.

13.5 Parental Selection

Parental selection is an essential part of the hybridization process. Starting with the appropriate parents can save time and money in developing a successful rose. In addition to those listed in *Modern Roses 12*, over two hundred rose species exist today, with approximately 10% used in breeding. As a result, the opportunity to create new roses is significant. However, combining all desired elements into one requires a hybridizer's ability to merge the efficiency of a scientist and the creativity of an artist, along with persistence and optimism.

Due to the complexity of gene inheritance and expression in roses, many unpredictable

Table 13.1. Example of a contemporary rose breeding scheme for California, USA. Author's own table.

Stage	Month	Tasks
1	January to March	Set up and prepare parent plants
2	April to July	Cross pollinations
3	October	Harvest and sorting hips
4	November to December	Stratification of hybrid seeds
5	January	Sow seeds by parentage
6	May to December	Primary positive selections
7	June to October	Rooting elite's selections
8	April to June	Distribute seedlings to global trials.
9	June to September	Propagation schedules
10	January to December	Marketing application for trademarks, registration, and patent rights

factors remain. To establish a solid foundation, it is important to focus on a specific goal and study the characteristics of each pedigree of selected varieties carefully. Backcrossing may be used to recover desired recessive genes.

Learning the strengths and weaknesses of various roses is a part of the rose breeder's job. Search for new blood to enrich the gene pool and continued vertical breeding from seedlings should be highly emphasized. Hybridizers are often stuck in their research by a "*seedling x seedling*" cross or are confused by unknown or doubtful parentage. *Modern Roses 12* is a good reference book to begin with. The Internet also provides particularly good informational access to *Rosa* family websites and facilitates further contact with rose enthusiasts and friends all over the world. Exchanges of seeds and pollen are quite commonly initiated online. Rose catalogs from rose suppliers and the *Combined Rose List of Dobson and Schneider* (Schneider, 2024) are other useful sources of information.

Deciding which parents to incorporate into a new cross is the first step to study. Native *Rosa* species and most healthy modern roses carry vigorous genes and/or horticultural traits.

Pollen viability and female fertility are two essential factors to seed setting. Pinching practice may be needed to control the timing of a pollination schedule, to synchronize the blooming time of both parents. And also, since male and female heritage are equally distributed to descendants, it is advised to select the better seed set variety as the leading female, and the scarce seed set variety as the male.

The closer the background of both female and male parents' pedigrees, the easier it is to accumulate their similar characteristics by backcrossing. However, the downsides to this "inline breeding" are lesser variation, lost vigor, and increased weakness, thus hybrid crossing produces greater vigor and greater variation. A wider cross induces more variation, but it takes a much longer time to recoup all desired characteristics.

A "perfect match" comes not only from luck but also from the painstaking observations of a careful strategy. It makes sense to pick only a few promising parents, or else start a long-term plan to accumulate the desired bloodline genes in the hopes that the created seedlings recoup enough desired genes to replace all the weaknesses or generate an interesting new feature.

It is advised to study the ploidy (e.g., chromosome number) compatibility while planning a cross, since most modern roses are tetraploid (4n), and the *Rosa* species are diploid (2n). *Rosa* species are tough, rough, never repeat once-in-a-season bloomers, and seem to carry wild and lesser-horticultural elements than modern hybrid roses. It might take generations of back-and-forth crossings to extract its horticultural value, assuming this procedure is not halted by a "mule" sterile or triploid plant (3n).

Another method to build up a gene pool is to acquire pollen from different regions of the world, where unique genes are available. This can be facilitated by express mail delivery services. However, there may be restrictions on sending or receiving pollen from certain countries. Hybridizers should contact fellow hybridizers via email to determine any applicable restrictive measures.

13.6 Pollination and Seed Management

From April to June in Southern California, when day temperatures range from 65–75°F (18–24°C), is a suitable time to start hand pollination. Anthers should be removed from the flower (emasculation) prime to release pollen. Pollen should then be applied 1 or 2 times to the same matured stigma to ensure pollen was accepted.

Hip set rate, germination rate, and selection rate are all affected by the freshness of the pollen. The right timing should be considered for applying pollen to the stigma. Typically, each hip contains 1 to 20 seeds and has around 50% germination rate.

Botrytis disease, insects, mites, and other hip-eating animals need to be controlled to avoid aborting pollinated hips. It is advised to keep a good feeding routine that provides adequate nutrition levels for the hips to set and the plant to grow. All hips should be harvested at the ripen stage and before the plant re-enters dormancy. Seeds should be extracted from hips, sterilized with 5% bleach, and stratified in a 32–35°F (0–2°C) refrigerator for a minimum of 8 weeks.

Breeders should prioritize the experience gained from each crossbreed. Meticulous record-keeping can enhance the direction and focus of the breeding program. While unexpected outcomes may occur, early identification of desirable traits will expedite achieving objectives. All collected seeds should be categorized by parentage, labeled with the appropriate code, and sown in freshly prepared soil beds with sufficient space prior to early spring. It is advisable to cover the seeds with thin cloths to retain moisture and protect against rodents that may harm emerging seedlings. These seedlings should be diligently nurtured for three months until they reach the blooming stage.

The volume of crosses made is often limited by budget and the availability of pollen and hip sources. If a single male does not provide enough pollen on its own due to the timing of flowering, then similar pollen, e.g., coming from a flower of a similar color, could be grouped to obtain enough viable pollen on the target flower. Likewise, females can be grouped as well to provide more opportunities to make crosses with the same goal in mind while targeted females are not available to pollinate.

13.7 Germination and Rogue Rate

The germination rate has a big effect on the selection rate. Obviously, more seeds mean more chances to make a viable selection. Some females produce hundreds of seeds per hip, while others only set 1 or 2 seeds per hip. However, this might be offset by many more blooms per plant.

While germination is affected by the fertility of males and females, so is culturing environment as well as their genetic composition. Since good germination provides more opportunities for selection, a lower discard rate is more meaningful than a germination rate. A compromise should be made between the seed set, germination rate, and selection rate. Thus, the selection rate could be the most important aspect to consider when making further crosses.

13.8 Evaluation

After parental matching plans, evaluation is one of the most critical as well as costly parts of a breeding program. It might take up to 10 years of persistent investments for a favorable outcome.

Some genetic traits are easily expressed in the first generation (dominant), but others may be too hidden (recessive). To recoup those hidden traits, further backcrosses or selfing must be made. This process might take further generations to get the result.

It is advised to study the pedigree to follow the goals of the cross and keep an open mind for surprises that may have happened occasionally.

No synthetic chemicals should be applied during the seedling growing period, especially fungicides from day one of testing. Constant observation and note taking should be completed by the end of the season. One should start with negative selection to weed out weak seedlings to give room for positive selection among healthy seedlings. This "weed out" procedure could save time and energy eventually.

In the first stage, each seedling should be evaluated, and data (Figs 13.1 and 13.2) should be collected in the seedbed and the open field for a minimum of 3 years. Most seedlings start to bloom by June. The first phase selection of tiny seedlings is based on color, basal break, disease resistance, and horticultural genetic traits. As it matures, the secondary selection phase aims at its growing habit, hardiness, fragrance, aging, and repeating blooming cycles. Promising selections should be tagged and transplanted to the field for continued evaluation of adaptability and rooting and grading capacity. Important heritable traits, selection rate, and germination rate should be recorded. Evaluation with note taking should be performed as often as possible. Notes for records should be kept as accurate and simple as possible.

Ten cuttings of each selection should be propagated and followed for 3 years from the original selection. These should be planted in different climatic zones for 3–5 consecutive years, including commercial trails.

After years of internal evaluation, propagated progeny of premier selection should be sent out for further geographical trials with partnership network testing sites, and/or submitted to participate in internationally associated competitions (American Garden Rose Selections, American Rose Society contests, All Germany Rose Trials/ADR, The World Rose Convention of the World Federation of Rose Societies). Winning awards in such competitions provide stronger promotion strength.

Fig. 13.1. Field evaluation of rose breeding populations.

All-America Rose Selections — Season Score Sheet

( ) 1st Spring ( ) 1st Fall ( ) 2nd Spring ( ) 2nd Fall	Date					Date					Date					Date					Season Score				
	Excellent	*Very Good*	*Average*	*Fair*	*Poor*	*Excellent*	*Very Good*	*Average*	*Fair*	*Poor*	*Excellent*	*Very Good*	*Average*	*Fair*	*Poor*	*Excellent*	*Very Good*	*Average*	*Fair*	*Poor*	*Excellent*	*Very Good*	*Average*	*Fair*	*Poor*
Novelty																									
Form, Buds																									
Form, Flowers																									
Color, Buds																									
Color, Open Blooms																									
Aging Quality																									
Flowering Effect																									
Fragrance																									
Stem/Cluster																									
Plant Habit																									
Vigor																									
Foliage																									
Disease Resistance																									
Repeat Bloom																									
Personal Opinion																									
Total Checks																									
Score Factor	10	x8	x7	x6	x4	10	x8	x7	x6	x4	10	x8	x7	x6	x4	10	x8	x7	x6	x4	10	x8	x7	x6	x4
Subtotal Scores																									
TOTAL SCORE																									

ENTRY NO. ______________

Judge ______________

Black Spot:	Please Check
Resistant	
Very Resistant	
Very Susceptible	

Mildew:	
Resistant	
Very Resistant	
Very Susceptible	

Appropriate Classification for this entry:

____

Does this rose exhibit qualities of a different class? ______________
If so, under which class would you categorize it? ______________

Season Height ______________

Season Width ______________

Abbreviations to use when entry cannot be scored:

E — Own Entry
D — Plants Dead
X — Unfair to Score (poor plants)
NB — No Bloom
NP — No Plants

COMMENTS, OBSERVATIONS, RECOMMENDATIONS: ______________

Fig. 13.2. Sample of a data sheet for recording the traits of rose progenies.

Evaluation should focus on identifying plants with a combination of the following features: disease and insect-resistant, hardiness, fragrance, quick blooming cycle, prolific in blooms, easy to grow, easy to own root, compact, vigorous, novel, thornless, shade-tolerant, stress-resistant, long-lasting blooms, heat and drought tolerant, geo-adaptability, and vase life.

Production of rooted cuttings also needs to be evaluated for a high percentage of Grade 1 finished bush.

13.9 Propagation

Currently, the use of own-root cuttings is recognized as one of the most efficient methods for propagating selections. This technique not only saves time and reduces costs but also effectively eliminates rose mosaic viruses that may be transmitted through budding rootstock.

The best rooting time in Southern California is May to October when the plant is in active growing condition. Choose healthy semi-softwood stems that show flower buds, to ensure quality and ease of management of each 'liner'. Every cutting should have one node or be a minimum of two inches in length. Rooting hormones such as auxins can be applied. The cutting shears and cuttings should be sterilized in bleach before being inserted into 50% perlite and 50% peat mix and placed under a mist system for 2 weeks. Rooted liners cultivated in a greenhouse bench (Figs 13.3 and 13.4) then could be potted and transplanted for field growth after 10 weeks of rooting. These liners could be shipped out for trial in a pot or field. Tissue culture to speed up building up elite varieties is also suggestible for mass production.

Fig. 13.3. Rooted rose liners.

13.10 Registration and Patenting

As soon as a decision has been made to introduce a variety commercially, one should ensure that the creation is fully protected by plant patent law. This is an incentive reward for breeders to compensate for their investment and provide financial security. It takes more than a year to process a patent application. Any seedlings commercially released on the market after a year or 5 years overseas do not qualify for patent application. Registering and trademarking of brand names and tagging (Fig. 13.5). Also, it can provide further security for promotions. Four key criteria in acquiring a successful plant patent are 'novelty,' distinction, uniqueness, and 'stability.'

13.11 Conclusions

There are thousands of available cultivars produced each year around the world. There is no room for another rose that does not exhibit the characteristics of beauty, health, and repeat blooming. Modern rose breeding is striving to further the perfection that has to combine at least one or more of the following qualities: pest

Fig. 13.4. Bench cultivation of hybrid rose progenies.

or disease resistance, novelty of flower form, size, color (non-fading), long shelf life, faster reproductive cycle, aging quality, petal texture, thornless, fragrance, glossy foliage, hardiness, ease of propagation, resistance to heat, drought, salt, and wind, shipping tolerance, etc. All of this should be pursued with a diversified growing habit to adapt to a wide range of geographical variances with minimum use of energy. Some of the novel cultivars bred by us fulfill above mentioned traits and are released to the market (Fig. 13.6A-H)

It takes up to 10 years of investment to combine such genetic puzzles. Large-scale commercial lead hybridizers see funding budget and teamwork as critical to launching new varieties in the market. Indeed, monetary benefits from royalties are big incentives to succeed. Therefore, controlling the budget and utilizing evidence-based expertise is recommended. Conventional breeding is still a workable tool to exploit many unknown *Rosa* species. In addition, mutations

Fig. 13.5. Branding and tag filled with information

Fig. 13.6. Novel rose cultivars were released by our group. A) Love&Peace. B) True Passion. C) True Sincerity. D) True Love. E) Golden Dragon (The Lone Star Rose). F) True Delight. G) True Serenity. H) True Perfume.

combined with *in vitro* techniques also have tremendous potential to change one or a few characteristics of an otherwise outstanding cultivar. Applying molecular techniques in the breeding of *Rosa*, especially in the market sector, could save both time and funding (Debener *et al.*, 2004; Gudin, 2010; Byrne, 2015; Smulders *et al.*, 2019; Zhang *et al.*, 2024). Hopefully the integration of conventional breeding methods and molecular techniques will make the beauty of roses more accessible to households worldwide.

Acknowledgments

Photos should be credited to my daughter Dr Jana Lim, who kindly contributed them to this report. I also thank Professor Fure-Chyi Chen for editing.

References

Byrne, D.H. (2015) Advances in rose breeding and genetics in North America. *Acta Horticulturae* 1064, 89–98. DOI: 10.17660/ActaHortic.2015.1064.11

Debener, T., Linde, M., Dohm, A., Kaufmann, H. (2004). The utilization of molecular tools for rose breeding and genetics. *Acta Horticulturae* 630, 29–42.

Gudin, S. (2010) Rose: genetics and breeding. *Plant Breeding Review* 17, 159–189.

Schneider, P. (2024) *Combined Rose List 2024*. The International Rose Directory, Mantua, OH, United States

Smulders, M.J., Arens, P., Bourke, P.M., Debener, T., Linde, M. *et al.* (2019) In the name of the rose: a roadmap for rose research in the genome era. *Horticulture Research* 6, 65. DOI: 10.1038/s41438-019-0156-0

Young, M.A. (2007) *Modern Roses 12*. 12th edn. American Rose Society, Shreveport, LA, United States.

Zhang, Z., Yang, T., Liu, Y., Wu, S., Sun, H. *et al.* (2024) Haplotype-resolved genome assembly and resequencing provide insights into the origin and breeding of modern rose. *Nature Plants* 10, 1659–1671.

Index

Note: Page numbers in *italic* type refer to figures and in **bold** type refer to tables

Printed and bound by CPI Group (UK) Ltd, Croydon, CR0 4YY

24/08/2026

14947852-0007